Dynamic Simulation of the Chemical Looping Combustion Process

# Dynamic Simulation of the Chemical Looping Combustion Process

**Vom Promotionsausschuss der**

**Technischen Universität Hamburg**

zur Erlangung des akademischen Grades

Doktor-Ingenieur (Dr.-Ing.)

genehmigte Dissertation

von

Johannes Haus

aus

Augsburg

2020

**Bibliografische Information der Deutschen Nationalbibliothek**
Die Deutsche Nationalbibliothek verzeichnet diese Publikation in der Deutschen Nationalbibliografie; detaillierte bibliographische Daten sind im Internet über http://dnb.d-nb.de abrufbar.
1. Aufl. - Göttingen: Cuvillier, 2020
Zugl.: (TU) Hamburg, Univ., Diss., 2020

Gutachter: Prof. Dr. Stefan Heinrich, Prof. Dr. Schlüter
Prüfungsleitung: Prof. Dr. Reimund Horn

Tag der mündlichen Prüfung: 29.10.2020


Nonnenstieg 8, 37075 Göttingen
Telefon: 0551-54724-0
Telefax: 0551-54724-21
www.cuvillier.de


1. Auflage, 2020
Gedruckt auf umweltfreundlichem, säurefreiem Papier aus nachhaltiger Forstwirtschaft.

ISBN 978-3-7369-7335-0
eISBN 978-3-7369-6335-1

**Table of Contents**

1. Introduction ... 9
2. State of the art ... 13
2.1. Fundamentals and modeling of fluidized bed reactors ... 13
2.2. Chemical Looping Combustion ... 20
2.2.1. Process routes in Chemical Looping Combustion and Conversion ... 22
2.2.2. Performance parameters to evaluate Chemical Looping Combustion ... 26
2.2.3. Pilot plant operation ... 28
2.3. Reaction kinetics ... 31
2.4. Modeling of Chemical Looping Combustion ... 39
3. Methodology ... 43
3.1. Experimental pilot plant ... 43
3.1.1. Layout and dimensions of the CLC pilot plant ... 43
3.1.2. Materials used inside the CLC pilot plant ... 47
3.1.3. Operation of the CLC pilot plant ... 48
3.1.4. Evaluation of the performance of the CLC pilot plant ... 51
3.1.5. Molar flows inside the pilot plant ... 52
3.1.6. Inventory loss during operation ... 54
3.2. Experimental lab scale fluidized bed ... 55
3.2.1. Reactor setup ... 55
3.2.2. Experimental procedure in the laboratory reactor ... 56
3.2.3. Data evaluation in the laboratory reactor ... 58
Molar balance from gas concentration measurements ... 58
Fitting the experimental data to particle reaction models ... 59
3.3. Dynamic Flowsheet Simulation ... 62
3.3.1. Hydrodynamic modeling of fluidized beds ... 64
3.3.2. Reaction kinetics in fluidized bed ... 68
3.3.3. Calculation procedure for the fluidized bed reactors ... 71
3.3.4. Modeling of the dynamic behavior for the fluidized bed reactors ... 74
3.3.5. Cyclone unit ... 77
3.3.6. The loop seal unit ... 78
4. Results and Discussion ... 81
4.1. Operation of the CLC pilot plant with different fuels ... 81
4.1.1. Gas concentration measurements at AR and FR ... 81
Fine lignite injected on lower stage ... 81
Fine lignite injected into the upper stage ... 82

Coarse lignite injected on lower stage ........ 83
Coarse lignite injected into the upper stage ........ 84
Fine bituminous coal injected into the lower stage ........ 84
Fine bituminous coal injected into the upper stage ........ 84
Coarse bituminous coal injected into the lower stage ........ 85
Coarse bituminous coal injected into the upper stage ........ 86
Methane injected along with fluidization gas ........ 87
Coarse Biomass conversion during injection into the lower stage ........ 88
4.1.2. Performance parameters for the pilot plant operation ........ 89
Carbon capture efficiency/oxide oxygen efficiency ........ 89
4.1.3. Influence of the second stage on the gas conversion ........ 95
4.1.4. Dynamic operation of the pilot plant ........ 96
4.2. Experimental lab scale fluidized bed ........ 100
4.2.1. Gas concentration measurements ........ 100
Inert bed gasification ........ 100
Reactive bed gasification ........ 100
Inhibition of inert sand bed gasification by hydrogen and carbon monoxide ........ 101
Influence of the fuel particle size on reactive bed gasification ........ 102
Simultaneous $CO_2$ and steam gasification ........ 103
Gasification with either $CO_2$ or steam ........ 104
Gasification under iG-CLC conditions: simultaneous $CO_2$ and steam gasification ........ 106
Gasification under iG-CLC conditions: variation of $CO_2$ and steam concentrations ........ 107
Gasification under iG-CLC conditions: variation of operating velocity $u_R$ ........ 108
Gasification under iG-CLC conditions: temperature dependence of simultaneous $CO_2$ and steam gasification ........ 108
4.3. Dynamic Flowsheet Simulation results ........ 112
Gas composition and situation in the FR stages ........ 112
Movement of bed masses inside the system ........ 114
Conversion of the oxygen carrier ........ 116
5. Summary and Conclusions ........ 119
5.1. Experimental pilot plant ........ 119
5.2. Laboratory kinetic reactor ........ 120
5.3. Dynamic Flowsheet Simulation ........ 121
List of Abbreviations ........ 123
List of Symbols ........ 125
References ........ 127

# Danksagungen

Die vorliegende Doktorarbeit wurde von mir am Institut für Feststoffverfahrenstechnik im Rahmen meiner Tätigkeit als wissenschaftlicher Mitarbeiter angefertigt. Besonderer Dank gilt dabei dem Institutsleiter Prof. Dr.-Ing. habil. Stefan Heinrich, der die Koordination des Gesamtprojektes sowie meines Teilprojektes im Schwerpunktprogramm der DFG übernommen hatte. Zusätzlich ermöglichte er es mir zeitlich wie auch finanziell an allen wichtigen wissenschaftlichen Konferenzen teilzunehmen, die den Fortschritt der Arbeit förderten. Auch das von ihm geschaffene Umfeld am Institut mit Besuchen von Gastwissenschaftlern und der regen Austauschmöglichkeiten mit starken Kollegen war für das Arbeiten sehr förderlich.

Einen herzlichen Dank will ich an Herrn Prof. Dr.-Ing. Joachim Werther aussprechen. Seine Art die wissenschaftlichen Ergebnisse in aller Tiefe zu besprechen und an der Erstellung der wissenschaftlichen Publikationen mitzuwirken half mir die notwendige, präzise Arbeitsweise zu entwickeln.

Herrn Dr.-Ing. Ernst-Ulrich danke ich für die unkomplizierte fachliche Betreuung, die Evaluierung von Messergebnissen und die Möglichkeit Problematiken an den Anlagen schnell zu besprechen und zu lösen.

Mit Heiko Rohde und Frank Rimoschat stand mir ein hilfsbereites und fachlich kundiges Personal zur Seite, welches mir bei der nicht immer reibungslose Arbeit an den Experimentalanlagen mit Rat und Tat zur Seite stand.

Für mich vorteilhaft war, dass von den beiden Vorgängern Marvin Kramp und Andreas Thon funktionierende Versuchsanlagen geplant, gebaut und in Betrieb genommen wurden, so dass ich schnell zu Ergebnissen gelangen konnte.

Bedanken möchte ich mich hierbei noch herzlich bei den Kollegen, welche dazu beitrugen, eine freundschaftliche, schöne Zeit am Institut zu gestalten. Meine Studierenden in Abschlussarbeiten und als wissenschaftliche Hilfskräfte machten es möglich die Breite an Ergebnissen zu erzeugen.

Zum Schluss möchte ich mich noch zutiefst bei meiner ganzen Familie bedanken, welche auf meinen Weg zum Doktortitel mitfieberten und auch vorher durch die materielle und ideelle Unterstützung im Studium zu meinem Werdegang beigetragen haben. Das schönste war die Geburt meiner Tochter Emilia und die Hochzeit mit meiner wundervolle Frau Marta, welche auch in die Zeit der Arbeit fielen.

Heidelberg, 08.11.2020

# 1. Introduction

Extraordinary weather conditions are felt to happen in an ever-increasing frequency in modern times. In history one can find many exceptional climate situations, but statistically relevant changes in the commonness of those can be noticed. The research community and most parts of the political establishment agree that the emission of anthropogenic greenhouse gases is the main reason for the changing climate conditions. Greenhouse gases can be mainly traced back to the economic activities and development of humankind; these are among others generating heat and electricity by fossil fuels, transportation powered by combustion engines, agricultural activities and fertilizer use as well as land use change. The greenhouse effect describes basically that gases in the atmosphere, mainly carbon dioxide, methane, ozone and water vapor, reflect the heat, which is emitted from earth's surface. On the one hand, this effect raises the overall temperature to livable temperatures, on the other hand, an increased greenhouse effect by higher concentrations of the mentioned gases in the atmosphere can lift the surface temperature to problematic levels and is then called global warming. Today, nearly all countries on this planet, decided that they want to limit global warming to 2°C in the Paris Agreement [1].

To achieve the 2°C target, especially the industrialized countries in the EU, China, India and the US must take strong measures to reduce their emission of climate active gases [2]. Most reliable research points in a direction that the reduction of climate gases is not even enough, and Negative Emission Technologies (NETs) have to be applied on a big scale to keep the temperatures on the planet in reasonable limits [3].

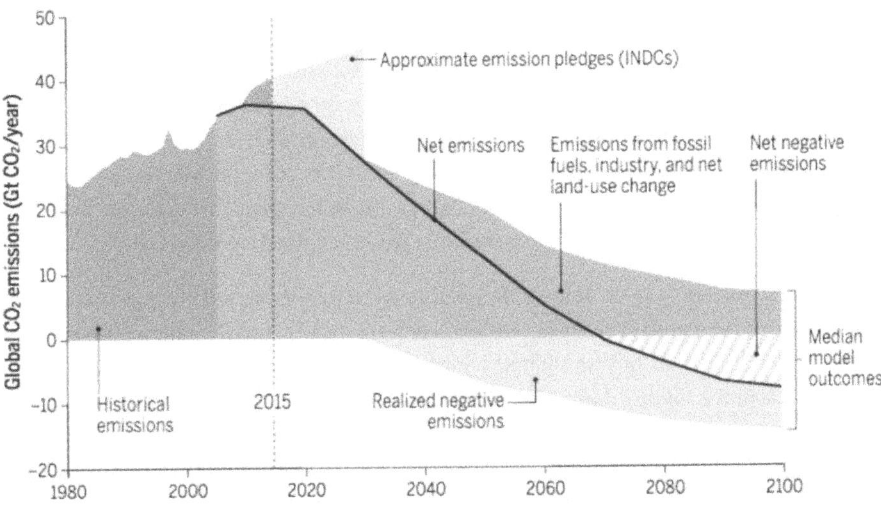


**Figure 1: A standard scenario of future $CO_2$ emissions to keep the 2°C global warming target. [4]**

For the energy sector this means that decarbonization of power production must take place. Decarbonization means that fossil fuel power plants are shut down in favor of renewable energies, like wind and solar. Heat can be generated by biomass, which is also considered mostly carbon neutral. Another approach to meet the climate goals are so called carbon capture and storage technologies (CCS). These technologies are used to capture carbon dioxide after the combustion process with the heat release. The carbon dioxide is then compressed and stored in various possible underground formations. CCS has the main advantage that the current energy system based on big shares of fossil fuel can be modified and does not need to undergo drastic changes. A major disadvantage of CCS technologies is still the elevated cost during electricity generation compared to venting the carbon dioxide into the atmosphere. This could be solved by carbon pricing and carbon trading, in which a reasonable price for carbon dioxide emissions is set by a governmental player and companies profit strongly from reducing them. Another reason why CCS could become important in the future is its ability to allow for negative emissions. When biomass is used in a thermal power station and the exhaust gases stored instead of emitted one can create a negative balance of emissions in the atmosphere, because the biomass absorbed carbon from the atmosphere before. In this case one speaks of Bio-Energy Carbon Capture and Storage (BECCS) [5].

Currently, some encouraging developments take place in the energy sector not only in Germany. Wind power turbines become economically competitive even during low electricity stock prices. Several days a year the complete electricity demand in Germany can be met by renewables. Contrariwise, when wind or sun generation is low, the need for power generation via thermal power plants arise. This is because the capacity of storing electricity is limited not only in Germany. If carbon dioxide emissions are to be avoided, CCS technologies are an option to provide energy in times of low renewable production.

One of the most discussed CCS technologies is the Chemical Looping Combustion (CLC) process. Contrary to the regular air-fired combustion, which takes place in one combustion chamber, the CLC process is carried out in two or more interconnected reactors. Between the reactors oxygen is transported via an oxygen carrying particulate, which is called oxygen carrier (OC) [6]. By separating the air intake to the process and the oxidation of the fuel in two different reactors, a pure stream of carbon dioxide can be achieved without the usage of air separation or flue gas carbon dioxide removal. The carbon dioxide should be relatively easily made available for compression and storage. The CLC process shows some interesting features, which make it a practical object for studies on heterogeneous reaction networks, interconnected fluidized bed reactors and the multiple involved reactive solids.

In the past, besides nuclear power plants, the energy system was based on fossil fuel fired power plants running on lignite and coal, which were erected to supply mid and base load. Peak loads were handled with power plants, which were a bit more flexible in their power supply, often gas turbines, hydro dams and pumped-storage hydroelectricity. Nowadays, the fluctuations inside the energy system are growing due to the intermittent renewables and existing regulating capacities are insufficient, leading to obscure phenomena, like negative or skyrocketing electricity prices. These effects are often attributed to the renewable energy use only, but another problem inside the energy system can be found in the thermal power plants itself. They cannot arrange their power output according to the demand fast enough due restrictions in load change rates, which are due to thermal expansion of the equipment and stress of the refractory lining [7]. Operation is most efficient when the designed power is achieved and, furthermore, personal talks of the author with plant operators and plant equipment suppliers revealed that there is limited knowledge about the operation of plants outside of the designed conditions, like shutdown and startup.

To bring renewable energies, fossil fuel fired power plants and novel thermal CCS technologies together in the future, the need for a more dynamic operation of the thermal plants arises, so that the intermittency of renewable sources can be accepted. This means that the thermal power plants will be operated often at part-load and changes in fuel load might be needed within minutes or hours. Also, shutdown and startup can be expected to happen more regularly if an energy system is achieved in which most of the electricity is provided by renewables.

For the design and implementation of flexible CCS thermal power plants, novel simulation tools could be beneficial, which can predict their operational behavior at a variety of conditions. One option for this is the widely used engineering tool of flowsheeting. In this approach, the modeling is broken down to all unit operations which constitute the entire process. To this day flowsheeting tools do not allow for dynamic analyses for complex solids processes. In this work a novel flowsheeting software tool called Dyssol is applied on a system of interconnected fluidized bed reactors used for CLC. A pilot scale and a lab scale reactor system are used to gather experimental data and reaction kinetics for an accurate flowsheet modeling of CLC.

This PhD thesis was financed and conducted within the framework of the research priority program of the German Research Foundation (DFG) SPP1679 "Dynamic Simulation of Interconnected Solids Processes". The goal of the research program is to develop simplistic and dynamic process models for a variety of solids processes. These models shall be integrated into a novel software, which is programmed in concourse of the entire research program.

The aim of this work is the development of process models for fluidized bed reactors and their integration into the simulation software Dyssol. Further, auxiliary equipment needed to model the exemplary process of Chemical Looping Combustion must be integrated into Dyssol. With the help of the dynamic models and the software, dynamic simulations have to be carried out, which will then be compared to experimental results at the pilot plant operated at the Institute. The pilot plant will be operated in a way to make dynamic effects, like start-up, shut-down and load changes visible. As a result, a validated software tool for dynamic simulation of interconnected fluidized bed reactors should be ready, which can later be used for process optimization and control of interconnected fluidized bed reactors.

Another aim of this work is to establish a database of experimental findings of the CLC process. This is done by searching for necessary series of experiments, which deliver the validation data for the simulations.

# 2. State of the art

According to scientific data based on various approaches, if humankind wants to achieve the 2°C goal put out during the Paris climate conference, strategies for CCS and NETs must be implemented into our energy systems [8]. Several concepts for the capture of climate gas emissions have been proposed. One of the major issues found in applying CCS technology is the nitrogen in the ambient air. Nitrogen has a fraction of around 79 vol.% in air and will be found in similar shares in the flue gases of regular air-fired boilers. For compression and storage of $CO_2$ this nitrogen creates a big overhead, which increases the costs for CCS to prohibitive levels. Also, $CO_2$ can be stored in its supercritical state due to the volume reduction, whereas this is not possible for $N_2$. Therefore, the most prominent technologies, namely Pre-Combustion, Oxyfuel und Post Combustion Capture, rely on an air separation process step, which heavily decreases overall efficiency of the process.

Another option is the Chemical Looping Combustion, where the separation of oxygen and nitrogen from the air takes place via a chemical reaction of a solid material. The research on CLC has reached several milestones in the meantime. The first being the research on suitable materials for the usage as OC in the system. A plethora of natural ores, refined ores, base chemicals and synthetic materials have been investigated for the suitability in CLC systems [9]. The second aspect is the successful operation of more than 15 CLC pilot plant units for solid fuels [9]. With the knowledge about OCs and pilot plant operation, process modeling was used to propose designs for upscaling considerations of the process to industrial scales. Standard terminology and evaluation procedures for the process were proposed in work of the research groups at Chalmers University of Technology in Gothenburg, Sweden and the Instituto de Carbochimica (CSIC-ICB) in Zaragoza, Spain, which resulted in review papers and overview works [10–13]. Song and Shen [14] reviewed most reactor concepts used in CLC.

In the present chapter, a process description is given in concourse with a definition of the main evaluation criteria for CLC. Afterwards, recent developments of the pilot plant operations in the years 2012-2018 are shown and concluded. This is because the two previous PhD theses by Thon [15] and Kramp [16] at Hamburg University of Technology (TUHH) were summarizing the field until 2012. This helps to discuss findings at the own pilot plant operation in Hamburg. Afterwards the status of process simulations for CLC is recapped to lead over to the own simulation work.

## 2.1. Fundamentals and modeling of fluidized bed reactors

The first CLC pilot operation was carried out in a system of interconnected fluidized bed reactors [6]. Since then most of the installed pilot plants were based on this technology. Two other reactor types have been reported, namely a moving bed reactor [17,18] and a rotating reactor [19], but they play a minor role in CLC research.

A fluidized bed is formed when an upwards directed flow through a particle bed exerts a force on the particles, which lifts them up and starts a movement of the particles. Depending on the shape and density of the particles and the porosity of the bulk as well as properties of the fluid used to fluidize, a certain velocity of the fluid is needed to enable fluidization, the so-called minimum fluidization velocity $u_{mf}$. If the reactor superficial velocity $u_R$ is above this velocity, sand-like or aeratable solids will form a fluidized bed. This was categorized early by Geldart, who investigated powders with different particle size distributions and densities for their fluidization behavior [20].

Depending on the operating superficial velocity $u_R$, different fluidization states of the particle bed can be achieved. Above $u_{mf,}$ particles will be put into motion and a bubbling fluidized bed (BFB will form, meaning that from the gas distributor bubbles will rise to the surface of bed. If the velocity is strongly increased, more and more particles will be entrained with the gas flow and no clear bed surface is visible anymore. At these conditions, a large mass flow of particles is moved out of the fluidized bed to a cyclone, where it is separated and led back to the lower part of the fluidized bed. This operation mode is called circulating fluidized bed (CFB), which shows other fluid mechanic structures than the bubbling bed, for example strands and clusters.

If the velocity is further increased, the exerted force on the particles is so high that they are directly lifted to the exit of the reactor. This setup is then called an entrained flow reactor, which is based on the pneumatic transport of all inserted particles. Distinctive operation states of fluidized beds are shown in Figure 2.

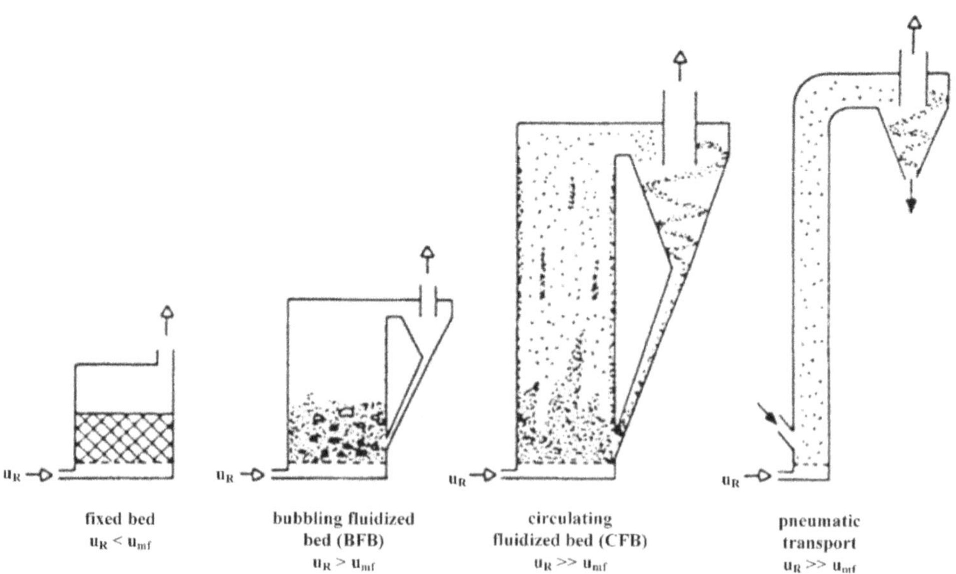


**Figure 2: Fluidized bed operation states depending on the reactor superficial velocity; translated from Wirth [21].**

The fluidized bed reactor concept was applied to a variety of processes within chemical and energy engineering. Major processes carried out are for example biomass and waste combustion, drying, fluid-catalytic cracking, agglomeration and coating. This is due to several advantages, which the fluidization of particles establishes. Those are for example the easy handling of solids material and the intense mixing of gases and solids. Further, in a fluidized bed a uniform temperature distribution can be achieved due to the large solid holdup combined with resolute movement of solids. During operation, high heat transfer coefficients between gas and particles as well as bed and internals and walls can be found compared to other reactor concepts. Altogether, this leads to a uniform product quality if the process is carried out properly.

Some disadvantages are usually connected with the usage of fluidized beds. The first one is the possibility for bed agglomeration and a subsequent defluidization of the particle bed. This can be caused for example by ashes during coal combustion, which have a relatively low melting temperature and then can form agglomerates with the bed material. Another aspect to mention is the scale-up from laboratory scale to industrial scale units, because small and local phenomena, like wall effects, can have tremendous effects on fluid mechanics. Further, the fluidization gas elutriates particles, which must be separated by cyclones or filters from the exhaust gases. Especially at high temperatures, the erosion induced by the particle flows can cause problems and the need for continuous maintenance. As a last disadvantage, it can be observed that the used particles have a broad residence time distribution meaning that for example with coal combustion some coal particles can leave the reactor unconverted.

The complex flow structure inside fluidized beds was the object of major research dedicated to describing and predicting the operation of fluidized beds. According to Werther et al. [22], CFBs are simulated on different length and timescales. For example, the Navier-Stokes equations are used in Computational Fluid Dynamics (CFD) code to resolve even the smallest scales of particle-gas suspensions on a 2- or 3-dimensional grid. Until now, CFD calculations of multiphase flows need considerable computational time to calculate only a small timeframe of several tens of seconds of operation.

In contrast to that, mathematical process simulation tools, like flowsheeting, are used to quickly estimate the macroscopic effects taking place in individual units of a complex process. Mathematical modeling can be based on black-box models, which simply solve mass and energy balances according to empirical or semi-empirical model equations or simple reactor models. These black box models can be enhanced with information about fluid dynamics of solids and gases as well as heat and mass transfer considerations to be able to describe the activity inside the process units more accurately. One approach to describe the hydrodynamics in fluidized bed reactors is 1-D modeling, which was used by numerous researchers summarized in Table 1. In this approach, the solids concentration of CFBs and BFBs is determined along the reactor height. Empirical and semi-empirical correlations extracted from a variety of operation conditions and fluidized bed geometries are available in literature. Further, main flow structures like bubbles were investigated and size correlations derived. In this work such empirical and semi-empirical models were used for the characterization of the fluidized behavior.

Modeling of bubbling bed reactors usually is based on a two zones approach, dividing the reactor volume in a dense lower zone and a dilute upper zone. The dense zone is characterized by the presences of a bubble and a suspension phase, which comprises a two-phase model. Sometimes a so-called wake phase with a lower solids concentration around the bubbles is included in the modeling. As explained above the bubbling bed will form a distinguishable surface, whose height can be determined by an overflow weir or a standpipe. Above that dense phase, a fraction of particles is entrained by the gas flow. Figure 3 shows an idealized bubbling fluidized bed with the bubble and suspension phase as well as the declining number of entrained particles above the dense bed. On the other side of the Figure, typical results of 1-D modeling are shown, which can be seen for example in Abad et al. [23] and Puettmann et al. [24]. This approach can be seen as the state of the art for the mathematical process modeling of bubbling fluidized beds. Differences between various models lie in different correlations, which are for example used for the bubble size development, the bubble rise velocity and the entrainment from the dense bed. These basic concepts can for example be found in the book "Fluidization Engineering" by Kunii and Levenspiel [25] or in the Handbook of Fluidization and Fluid-Particle Systems edited Yang [26].

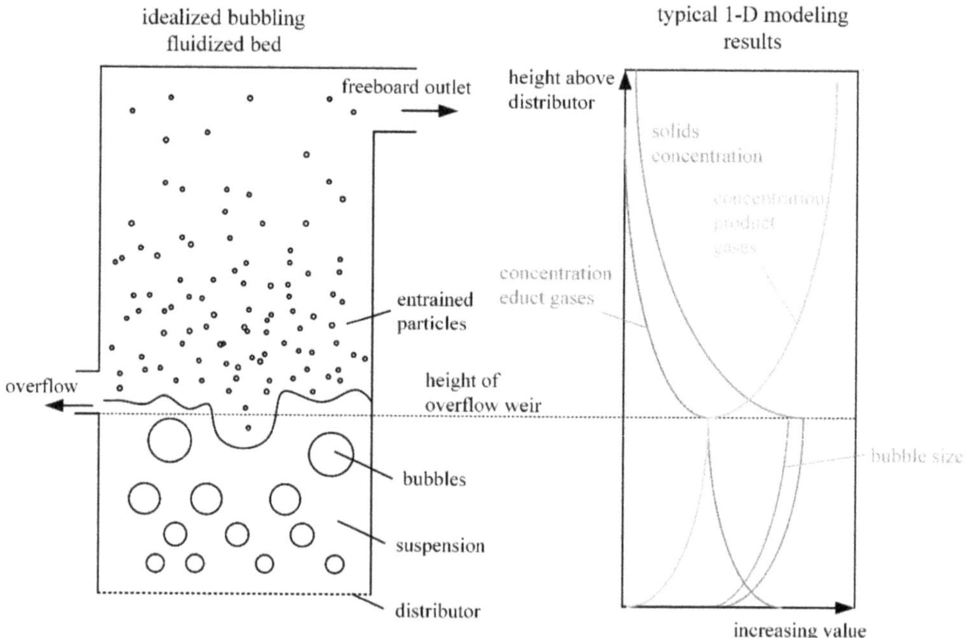


**Figure 3: Idealized visualization of a bubbling fluidized bed and typical outcome of 1-D modeling of bubbling fluidized beds.**

The flow structure and the physical phenomena in CFB riser reactors differ from what was seen in bubbling beds. Even though several researchers found a dense bed at the bottom of their riser reactor, the solids concentration there is typically only a fraction of the one seen in bubbling beds. This is due to the presence of different meso structures, like strands and clusters, in CFB risers, which are more solid lean. On the other hand, towards the top of the reactor, usually one can find higher solids concentrations than the ones seen in bubbling beds, due to the higher drag force, which is induced on the entrained particles. Furthermore, there is a radial gradient of solids concentration in CFB risers, which is traced back to the presence of a core-annulus flow structure. In the core of the reactor, particles are pneumatically transported upwards at high velocities. In an annulus close to the wall, the upwards velocity becomes much smaller or even negative and, with it, the particles are transported downwards there. These flow phenomena were investigated broadly by a series of researchers, which investigated the axial 1-D distribution of solids in CFB risers, the size of the core-annulus flow, or even the radial distribution of solids at certain reactor heights. Major findings are summarized in Table 1 and arranged according to their investigation aim.

**Table 1: Modeling of CFB riser reactors. Axial 1-D modeling, annulus size determination and radial resolution of solids position in reactors.**

| researcher | approach/ focus | description/governing equations | ref. |
|---|---|---|---|
| Li and Kwauk (1980) | 1-D | Porosity calculated from known porosity of lower bed and at the end of the riser. Knowledge about inflection point needed to plot porosity profile.<br>$\ln\frac{\varepsilon-\epsilon_a}{\varepsilon^*-\varepsilon} = -\frac{1}{Z_o}(z - z_i)$<br>zi = position of inflection point<br>$\varepsilon^*$ = porosity lower bed<br>$\varepsilon_a$ = porosity riser exit<br>$Z_0$ = characteristic length of fast fluidization | In book: [27] |
| Kunii and Levenspiel (2014) | 1-D dense bottom, dilute upper zone | A dense bottom zone with known solids concentration is used together with the assumption of an exponential decay of solids concentration above the dense phase. knowledge about dense phase porosity needed.<br>Dilute upper zone is calculated via:<br>$\frac{\varepsilon_s-\epsilon_s^*}{\varepsilon_{sd}-\epsilon_s^*} = e^{-az_f}$<br>The dense bottom is considered constant at $\varepsilon_{sd}$.<br>$z_f$ = height above distributor<br>$\varepsilon_s$ = porosity at height $z_f$<br>$\varepsilon_s$* = saturation porosity riser exit<br>$\varepsilon_{sd}$ = porosity bottom riser<br>a = decay constant for exponential elutriation | [25] |
| Zhang et al. (1981) | 1-D dense bottom, dilute upper zone | The riser is modeled first with an increase of solids concentration from a known starting value. From a known height of a dense bed, an exponential decay towards a known outlet concentration is calculated.<br>Dilute upper zone is calculated via:<br>$\varepsilon = \varepsilon_a + \frac{\epsilon^*-\epsilon_a}{2} e^{\frac{-(z-z_i)}{A}}$<br>The dense dense bottom zone follows:<br>$\varepsilon = \varepsilon^* - \frac{\epsilon^*-\epsilon_a}{2} e^{\frac{(z-z_i)}{A}}$<br>A = characteristic length<br>$\varepsilon^*$ = porosity lower bed<br>$\varepsilon_a$ = porosity riser exit<br>z = height of dense bed | In book: [28] |

**Table 1 (continued)**

| | | | |
|---|---|---|---|
| Bai and Kato (1999) | 1-D dense bottom, dilute upper zone | 17 sets of experimental data investigated about their solids concentration along riser height.<br>The proposed model equations distinguish between dense bottom zone and dilute upper region. For both regions 2 sets of correlations were proposed. They separate the flow regimes according to the solids circulation $G_s$ and whether it is above or below a so-called saturation carrying capacity $G_s^*$.<br>Dilute upper zone is calculated via:<br>$\frac{\varepsilon^*}{\epsilon_s'} = 4.04\, \epsilon_s'^{0.214}$ for $(G_s < G_s^*)$<br>$\frac{\varepsilon^*}{\epsilon_s'} = 1 + 0.208\left(\frac{G_s}{\rho_s U_0}\right)^{-0.5}\left(\frac{\rho_s-\rho_g}{\rho_g}\right)^{-0.082}$ for $(G_s > G_s^*)$<br>The dense bottom zone is described as:<br>$\frac{\varepsilon_{sd}}{\varepsilon_s'} = 1 + 0.00614\left(\frac{G_s}{\rho_s U_0}\right)^{0.23}\left(\frac{\rho_s-\rho_g}{\rho_g}\right)^{1.21}\left(\frac{\sqrt{gD}}{U_0}\right)^{0.383}$<br>for $(G_s < G_s^*)$<br>$\frac{\varepsilon_{sd}}{\varepsilon_s'} = 1 + 0.103\left(\frac{G_s}{\rho_s U_0}\right)^{-1.13}\left(\frac{\rho_s-\rho_g}{\rho_g}\right)^{-0.013}$<br>for $(G_s > G_s^*)$<br>An ideal solids volume concentration is assumed at:<br>$\varepsilon_s' = \frac{G_s}{\rho_s(U_0-V_t)}$<br>$\varepsilon_{sd}$ and $\varepsilon^*$ = upper and lower limit of solids concentration at bottom and top of reactor, respectively<br>$G_s^*$= saturation carrying capacity of gas | [29] |
| Werther and Hartge (2004) | 1-D dense bottom, dilute upper zone | A dense bottom zone is modeled with a bubble and a suspension phase. The governing equation is the bubble size change above the distributor. The dilute region above the dense region is characterized by an exponential decay of solids concentration.<br>dilute upper phase:<br>$c_{v,i}(h) = \left(c_{v,d,i} - c_{v,i}^*\right)e^{-(a(h-H_d))} + c_{v,i}^*$<br>dense bottom zone:<br><br>$c_v$ = solids concentration<br>h = discretized height, $H_d$ = height of dense zone<br>i = particle size class<br>a = decay constant<br>$d_v$ = bubble size<br>$\varepsilon_b$ = bubble volume fraction<br>$\lambda$ = bubble lifetime | [30] |
| Werther (2002) | Annulus size | For radial profiles, the annulus thickness needs to be known. The research investigated large scale units for the annulus size and defined the empirical correlation<br>$\left(\frac{\delta}{D_t}\right) = 0.55\, Re_t^{-0.22}\left(\frac{H_t}{D_t}\right)\left(\frac{H_t-z}{H_t}\right)$<br>$\delta$ = diameter between core and annulus<br>$D_t$ = diameter of the reactor<br>$H_t$ = Total height of the reactor<br>z = height above distributor<br>Re = Reyonolds number based on reactor diameter | In book: [31] |

**Table 1 (continued)**

| | | | |
|---|---|---|---|
| Patience and Chaouki (1993) | Annulus size | Regression over a large amount of experimental data gave the correlation of the annulus size.<br>$\Phi_g = \left(\frac{r_c}{R}\right) = \frac{1}{1+1.1\,Fr\left(\frac{G_s}{\rho_s U_0}\right)^{0.083Fr}}$<br>$r_c$ = diameter between core and annulus<br>Fr = Froude number<br>$G_s$ = solids circulation<br>$U_o$ = gas velocity | [32] |
| Nakamura and Capes (1976) | Annulus size | With help of several parameters and measured voidage profiles of the core and the annulus, the annulus size can be calculated at any point above the distributor via:<br>$1-\left(\frac{r_c}{R}\right)^2 = \frac{\sqrt{\lambda(1-\varepsilon_c)(1-\varepsilon_a)\left(-\frac{U_0}{V_t}+\varepsilon_c^n+\frac{\frac{\varepsilon_c V_p}{V_t}}{1-\varepsilon_c}\right)}}{\varepsilon_c-\varepsilon_a}$<br>$r_c$ = core radius<br>R = Riser diameter<br>λ = parameter determined by terminal velocity and riser diameter<br>n = parameter<br>$\varepsilon_c$ = average core voidage<br>$\varepsilon_a$ = average annulus voidage<br>Vt = terminal velocity<br>$U_o$ = gas velocity | [33] |
| Issangya et al. (2001) | radial distribution | By knowing the porosity at minimum fluidization and the axial porosity, the radial distribution of solids at any height can be calculated:<br>$\varepsilon(r,z) = \varepsilon_{mf} + (\varepsilon(z) - \varepsilon_{mf})\varepsilon(z)^{-1.5+2.1\left(\frac{r}{R}\right)^{3.1}+5\left(\frac{r}{R}\right)^{8.8}}$<br>r = radial position<br>R = riser diameter<br>ε(z) = axial average porosity at a certain height z<br>$\varepsilon_{mf}$ = voidage at minimum fluidization | [34] |
| Zhang, Tung and Johnsson (1991) | radial distribution | The model makes it possible to describe the radial porosity and with it the solids distribution if the average porosity at a certain height is known.<br>$\varepsilon(r) = \varepsilon^{0.191+\left(\frac{r}{R}\right)^{2.5}+3\left(\frac{r}{R}\right)^{11}}$<br>r = radial position<br>R = riser diameter<br>ε = axial average porosity at a certain height | [35] |

**Table 1 (continued)**

| | | | |
|---|---|---|---|
| Pugsley and Berruti (1996) | 1-D with radial distribution | The model divides the riser reactor in an acceleration region at the bottom and a fully developed region above it. First the slip velocity between particles and the fluidization gas is calculated. Then porosity in the fully developed region is determined. In a next step the acceleration region is determined, by calculating how long the particles need to be accelerated to the final particle velocity found in the fully developed region.<br>For the fully developed region it was follows:<br>$\Psi = \frac{U_0}{\varepsilon V_p} = 1 + \frac{5.6}{Fr^2} + 0.47 Fr_t^{0.41}$<br>$\varepsilon = \frac{U_0 \rho_s}{G_s \Psi + U_0 \rho_s}$<br>In the acceleration region it was proposed:<br>$\frac{dV'_{p,c}}{dt} = \frac{3}{4} C_d \frac{\rho_g V_{sl}^2}{d_p \rho_s} + \frac{g(\rho_g + \rho_s)}{\rho_s}$<br>$\Psi$ = slip velocity between particles and gas<br>$U_0$ = gas superficial velocity<br>$V_p$ = particle velocity<br>Fr = Froude number<br>$Fr_t$ = Froude number based on terminal velocity<br>$\varepsilon$ = porosity in the riser<br>$C_d$ = drag coefficient | [36] |

## 2.2. Chemical Looping Combustion

In a patent from 1954, when climate change was not yet examined and discussed, Lewis and Gilliland [37] proposed a chemical reactor system, which produces a pure steam of $CO_2$. The issue that they worked on was the need for pure $CO_2$ for Enhanced Oil Recovery (EOR). Their patent describes an interconnected system of fluidized reactors, in which a ferrous solid oxygen carrier transports the oxygen needed for the conversion of the fuel. Their intent was to prevent the dilution of flue gases with nitrogen, which happens during regular air-fired combustion.

Ishida et al. [38] concluded that the looping of a solid oxygen carrier (OC) could improve the plant efficiency of a liquefied natural gas plant. The oxidation and reduction reaction of several metal oxide carriers were investigated about their energetic potential for usage in a two-reactor system. In further work, the reduction and oxidation reactions of several of those metal oxides were investigated in a thermogravimetric analyzer (TGA) [39]. Also different fuel gases were introduced to the metal oxides and further it was checked, how the nitrogen oxide formation looks like under those operation conditions [40]. Already in their publications from the late 1980s, it could be shown that CLC has the potential for a low NOx generation during the fuel conversion, which can be traced back to the absence of nitrogen.

The entire process concept was refined and put into place in a real reactor first time by Lyngfelt et al. [6] in 2001. The research group built a dual fluidized bed test facility and could establish a CLC operation with methane as fuel. In this publication, they put out a basic process schematic, which is shown in Figure 4, and has since then used steadily by the CLC research community. As one can see there, the general layout consists of two interconnected reactors, an air reactor (AR) and a fuel reactor (FR). Between both reactors, a solid OC is circulated. The OC consists usually of a metal oxide MeO, which is oxidized in the air reactor and reduced in the fuel reactor. Also, other carriers based on for example $CaSO_4$ [41] were proposed. The different metal oxides differ sharply in their thermodynamic reaction behavior towards the fuel gases. Some carrier materials can chemically release gaseous oxygen in the FR, if temperature and pressure conditions are suitable. This effect is called Chemical Looping with Oxygen Uncoupling [42].

It is shown in the Figure that in the best case, the AR exhaust gas consists of mainly nitrogen and some unused oxygen, whereas in the FR a pure stream of $CO_2$ and steam could be exhausted. To achieve this separation of both reactors' exhaust streams, a good gas sealing between the reactors needs to be established. If the process is carried out properly, the CLC process offers a unique option to reduce the efficiency penalty that is normally induced by other carbon capture processes due to the air separation units [43].

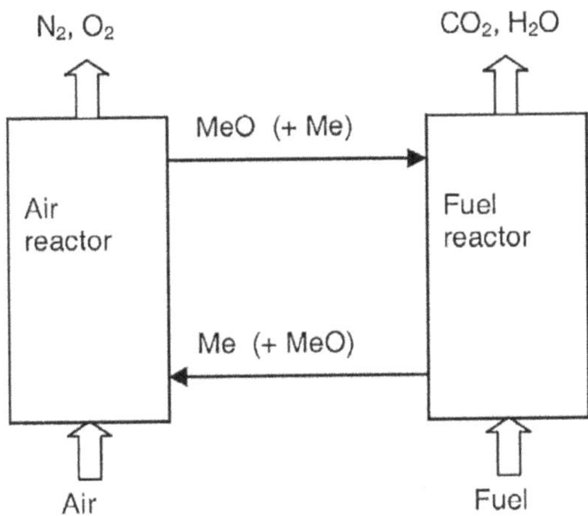


**Figure 4: Reactor concept as proposed by Lyngfelt et al. [6]**

Lyngfelt et al. [6] used a system of interconnected fluidized bed reactors for carrying out the CLC process. Their system is shown in Figure 5, where a circulating fluidized bed riser AR is connected to a bubbling fluidized bed FR. A cyclone separates the OC from the AR flue gases from the circulating OC. Two loop seals in between the reactors are there to prevent gas flows from the FR and the AR and vice versa.

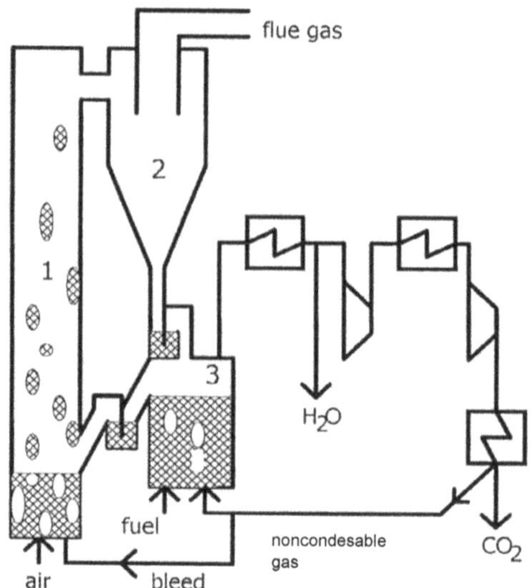


**Figure 5: Dual fluidized bed operated by Lyngfelt et al. [6]. 1 = air reactor, 2 = cyclone, 3 = fuel reactor.**

### 2.2.1. Process routes in Chemical Looping Combustion and Conversion

For demonstration purposes, the first deployed CLC reactor was running on the gaseous fuel methane [6]. Unfortunately, this fuel is unsuitable for industrial scale CLC units since more efficient heat and power processes are available for this fuel, namely the Natural Gas Combined Cycle (NGCC). This latter process reaches electrical efficiencies of up to 60% of the thermal input power, which cannot be reached with a CLC reactor. The high efficiency allows for other carbon capture methods, like post-combustion capture with amines, to be deployed and at the same time reaching higher overall efficiencies [44]. This is the reason why current research focuses mainly on the conversion of solid fuels with the CLC process. Liquid fuels have also been investigated for the use in a CLC setting.

The process routes have been described in detail by Adanez et al. [10] and Lyngfelt et al. [9], but a small summary will be provided in this chapter. There are three major routes for CLC proposed, those are in-situ Gasification CLC (iG-CLC), Chemical Looping with Oxygen Uncoupling (CLOU) and Chemical Looping Reforming (CLR). They differ mainly by the way the fuel is converted in the system and the kind of oxygen carrier used. Due to high temperature and mechanical attrition inside the system, elutriated fines of the OC leave the reactor constantly and have to be made up for. This leads to economic considerations, which carrier material should be used in the system, to minimize the cost of make-up material. Usually, the lower cost OCs based on natural ores, which are directly used as bed materials, have worse reactivity to convert the fuel gases.

<u>In-situ Gasification Chemical Looping Combustion (iG-CLC)</u>

The first process route is the in-situ Gasification CLC, in which the solid fuel is directly fed into the FR. Due to the high temperatures inside the reactor, the volatile fraction of the solid fuel is released fast and can react directly with the solid OC. The remaining char needs to be gasified by $CO_2$ and steam, which is present as a fluidization gas, but also is a conversion product of the reaction of the OC with the volatiles. Figure 6 illustrates the conversion mechanism of solid fuels in an iG-CLC plant. The chemical reactions are written in a general form in equations according to Adanez et al. [10]. Equations (1) and (2) describe the gasification of solid carbon char in the FR with the gases steam and/or carbon dioxide. The gasification products as well as the volatiles can react with the metal oxide OC as described in equation (3). In the gas phase, the homogeneous water-gas-shift (WGS) reaction, equation (4) can take place. The depleted OC must be reoxidized according to equation (5) in the AR.

| | | |
|---|---|---|
| steam gasification: | $C^s + H_2O^g \rightarrow H_2^g + CO^g$ | (1) |
| gasification with $CO_2$: | $C^s + CO_2^g \rightarrow 2\ CO^g$ | (2) |
| gas conversion | $H_2^g$, $CO^g$, volatiles$^g$ + n $Me_xO_y^s \rightarrow CO_2^g + H_2O^g$ + n $Me_xO_{y-1}^s$ | (3) |
| water-gas-shift: | $H_2O^g + CO^g \rightarrow H_2^g + CO_2^g$ | (4) |
| reoxidiation: | $Me_xO_{y-1}^s + \frac{1}{2} O_2^g \rightarrow Me_xO_y^s$ | (5) |

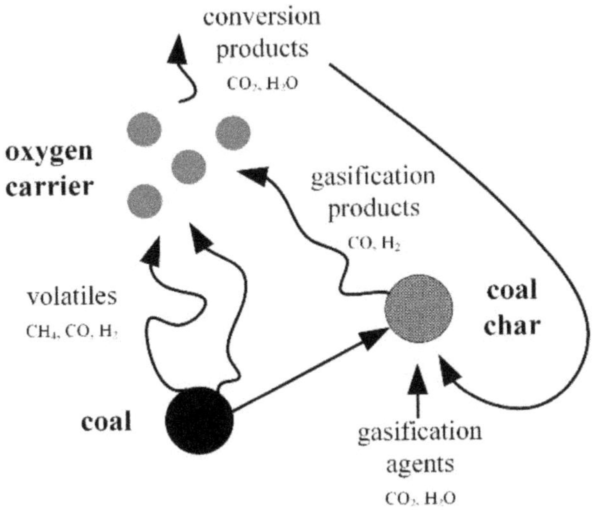


**Figure 6: Conversion of Coal in an iG-CLC plant adapted from Adanez et al. [10].**

The major advantage of iG-CLC is that the solid fuel can be directly fed to the FR. Further, usually cheap oxygen carriers are available for the conversion of the fuel gases, which are mainly $CH_4$, CO and $H_2$. Suitable oxygen carriers for the iG-CLC route have been reported to be based on iron oxides $Fe_2O_3$, nickel oxides NiO, copper oxides CuO, cobalt oxides $Co_3O_5$ and manganese oxides $Mn_2O_3$. Due to high temperature and mechanical attrition inside the reactor system, the nickel and cobalt base OCs pose an environmental risk due to the toxicity of the elements [45].

One major disadvantage of the iG-CLC process route is the need for the gasification of the solid fuel. In general, one can say that the gasification is radically slower compared to regular oxidation of the solid fuel with air. Even though it has been shown in own research that the gasification is enhanced by the presence of OC compared to an inert sand bed (cf. chapter 4.2), the gasification of the char is often a bottleneck in the efficient conversion of the solid fuel.

Chemical Looping with Oxygen Uncoupling (CLOU)

To overcome the limitation of the gasification rate of the solid fuel, the special ability to release gaseous oxygen by cobalt-based, manganese-based and copper-based OCs was demonstrated. Mattisson et al. [42] showed that at certain temperature levels, these metal oxides will release a certain amount of elementary oxygen according to a chemical equilibrium with the gas phase around the OC. The oxygen partial pressure that can be achieved at atmospheric operation pressure are visualized in Figure 7. CLOU operation changes the reaction network with the solid fuel completely as now the char of the solid fuel can react directly with the gaseous oxygen released. Adanez et al. [10] describe the prevailing reactions generally with equations (6) - (9).

| | | |
|---|---|---|
| oxygen release: | $2\ Me_xO_y^{\ s} \rightarrow 2\ Me_xO_{y-1}^{\ s} + O_2^{\ g}$ | (6) |
| solid fuel conversion: | $Coal^s \rightarrow Volatiles^g + Char^s + H_2O^g$ | (7) |
| char oxidation: | $Char^s + O_2^{\ g} \rightarrow CO_2^{\ g}$ | (8) |
| volatiles oxidation: | $Volatiles^{\ g} + O_2^{\ g} \rightarrow CO_2^{\ g} + H_2O^g$ | (9) |

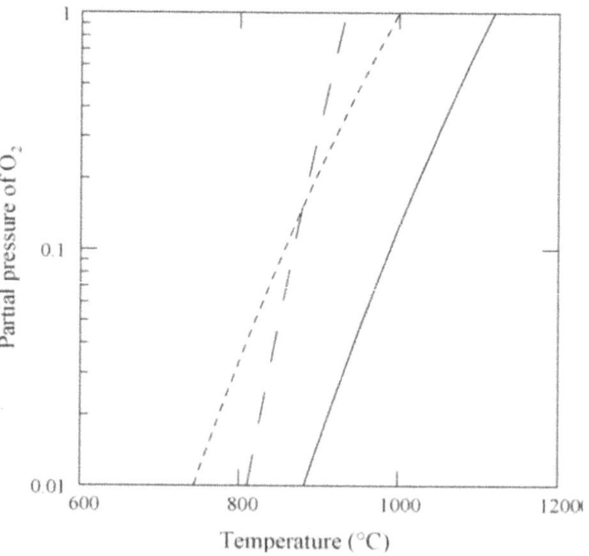


**Figure 7: The partial pressure of gas-phase O2 for the metal oxide systems $CuO/Cu_2O$(—), $Mn_2O_3/Mn_3O_4$ (- - - - -) and $Co_3O_4/CoO$ (— —) as a function of temperature from Mattisson et al. [42].**

Other Chemical Looping Processes: Chemical Looping Reforming (CLR)

There are several other process routes using the basic principle of Chemical Looping, the circulating of an OC for providing a certain amount of oxygen to a fuel. If the main purpose of the process layout is not the generation of heat and its use for electricity generation, one can operate a CLC system in a way to generate hydrogen from fuels and at the same time capturing connected carbon emissions. The produced hydrogen could be for example used for the decarbonization of the transport sector.

The simplest way under CLC conditions to produce a syngas, which later can be transformed to $H_2$, is to increase the fuel load in the FR. Also, the amount of oxygen, which is transported from the AR to the FR can be decreased. Both measures lead to a situation, where not enough oxygen is presented to fully oxidize the fuel gases and, hence, the off-gas contains a certain amount of syngas. This syngas can be reformed via the WGS reaction (4) towards higher concentrations of $H_2$ and $CO_2$. An additional adsorption step for the $CO_2$ and condensation of the steam can then create separated streams of $CO_2$ and $H_2$.

Ryden et al. [46] proposed the design of Figure 8, where a steam reforming plant is modified with a CLC system to supply the heat into a steam reformer reactor. The steam reformer is designed as a FR with internal heat transfer surfaces to transfer the heat to methane/steam mixtures. The heat for the steam reforming is mainly provided by the heat carried to this steam reformer reactor from the AR oxidation of the OC. This concept separates the steam reforming cycle and the regular CLC cycle, which provides the heat.

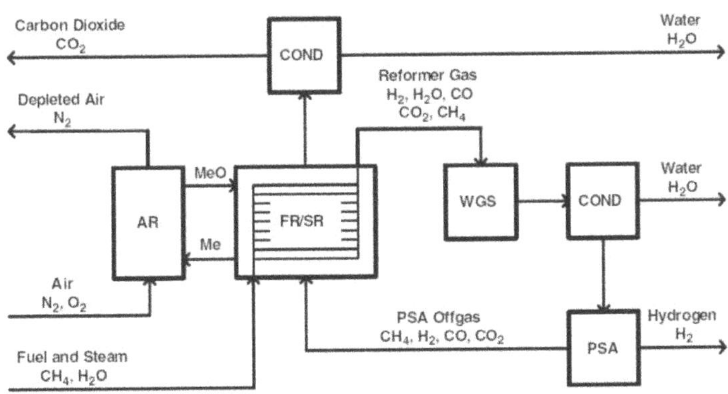


**Figure 8: Steam reforming plant with carbon capture from Ryden et al. [46]. (SR = steam reformer, WGS = water-gas-shift reactor, PSA = pressure swing adsorption).**

A less complex design depicted in Figure 9 was proposed by Ortiz et al. [47], which introduces only one fuel gas, which is used partly for the heat release needed for the steam reforming and partly as a precursor for the steam reforming itself. The process is controlled mainly by the oxygen introduced from the AR to the FR. The steam reforming is an endothermic process, which needs a certain amount of heat to be sustained. Hence, parts of the introduced fuel gas are oxidized to provide the needed process heat. The more oxygen provided to the FR, the more methane is converted to $CO_2$ and steam, the less syngas can be used for reforming and vice versa.

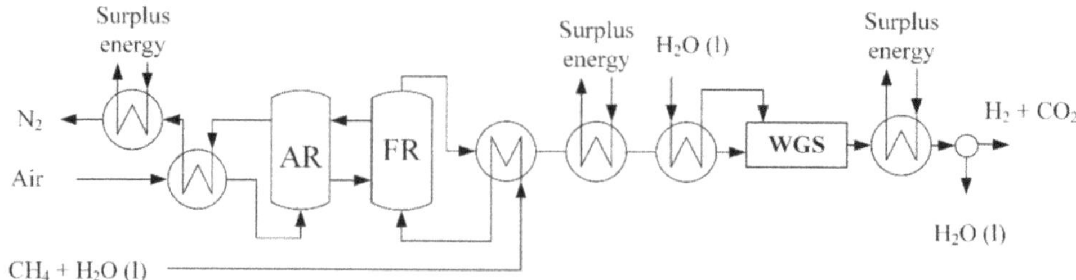


**Figure 9: Autothermal steam reforming process with carbon capture as proposed by Ortiz et al. [47].**

### 2.2.2. Performance parameters to evaluate Chemical Looping Combustion

Under optimal circumstances, the CLC process would provide the opportunity to obtain a pure stream of $CO_2$ during the conversion of fuels in thermal power plants, without adding high costs compared to regular air-fired combustors. Unfortunately, the operation of first pilot plants revealed several mechanisms, which have a negative impact on the overall process, which must be addressed, if the process is implemented into the energy system. To avoid later purification steps by achieving high efficiency of separation, the off-gas from the AR should contain at best only $N_2$ and some remaining $O_2$ and the off-gas from the FR should be $CO_2$ and $H_2O$ only.

Unfortunately, there are some unwanted gas flows from the AR to the FR and vice versa, which dilute the off-gases of both reactors. This is visualized in Figure 10. If the gas slips from the FR to the AR via the loop seals, $CO_2$ can be found in the AR off-gas and is then not captured in CLC operation. If gas slips the other way from AR to FR, the FR off-gas is diluted with $N_2$ and $O_2$, which makes the later compression and storage more expensive. It is not yet clear, which quality the off-gas of the FR needs to have regarding the purity of $CO_2$.

Another major issue during the operation of CLC plants is the unconverted fuel inside the FR. In the iG-CLC process route, it may happen that the solid fuel is not fully gasified until it is transported to the exit of the FR. Those solid fuel parts are usually referred to as carbon fines and will be found in the filters of the flue gas treatment. Another issue to mention is the gaseous fuel gases, mainly CO, $H_2$ and $CH_4$, which leave the FR unconverted. They originate as volatiles directly from the solid fuel or are generated during the gasification of the fuel with steam and $CO_2$. Parts of these fuel gases do for various reasons not contact the OC and, hence, are not oxidized towards the desired products $CO_2$ and steam. The more fuel leaves the reactor unconverted, the lower is logically the plant efficiency regarding the fuel input. Further, unconverted fuel could have negative effects on the storability of the flue gas.

If gasification of solid fuels in the FR is not fast enough then it is possible that solid fuel char is carried over from the FR to the AR. In the AR, a surplus of oxygen is present, which will usually completely convert the char towards CO2. The main aim of CLC is a high concentration of that gas in the FR product gas and to achieve a low concentration of CO2 in the AR off-gas. To realize a highly efficient process, the carbon slip must be kept as low as possible.

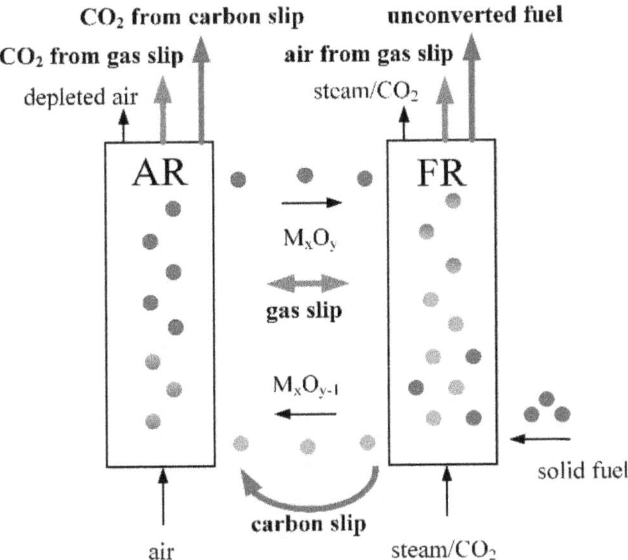


**Figure 10: CLC concept showing unwanted effects during the operation of a CLC plant.**

The described unwanted effects of carbon slip, gas slip and the exhaust of unconverted fuel parts are evaluated for a CLC system based on molar flows of in and outgoing streams. The effects find their expression in dimensionless parameters, the carbon capture rate $\eta_{CC}$ or $\eta_{OO}$, the fuel conversion $X_{SF}$and the oxygen demand $\Omega_{OD}$. The definition of those deviates slightly between research groups and, therefore in the experimental section a detailed molar balancing is presented. An overview over the presently used definitions can be taken from Adanez et al. [11]. In this section the definitions are described, which are used in the present work to be able to discuss the pilot plant operation of various research groups.

Carbon capture and oxide oxygen efficiency

The main purpose of the CLC process is to capture as much carbon dioxide as possible during the fuel conversion of both gaseous and solid fuels. This is expressed as the carbon capture efficiency $\eta_{CC}$ of a reactor system. The $\eta_{CC}$ relates the outgoing carbonaceous streams from both reactors, to the carbonaceous stream from the FR only, equation (10). If the rate is close to the value of 1, then most carbon leaves the system via the FR, where it is desired to leave. Lower values indicate that gaseous and/or solid carbon slip from the FR to the AR occurs.

$$\eta_{CC} = \frac{\dot{n}_{carbon,FR,out}}{\dot{n}_{carbon,FR,out} + \dot{n}_{carbon,AR,out}} \qquad (10)$$

Since the carbon streams can only be obtained by precise measurement of all in and outgoing carbon streams, often another indicator is used, the oxide-oxygen efficiency. This efficiency is based only on the flows inside the AR and can be determined easier according to equation (11). If measurements and mass flow are taken accurately, both efficiencies should yield same values.

$$\eta_{OO} = \frac{\dot{n}_{O_2,AR,in} - \dot{n}_{CO_2,AR,out} - \dot{n}_{O_2,AR,out}}{\dot{n}_{O_2,AR,in} - \dot{n}_{O_2,AR,out}} \tag{11}$$

Solid fuel conversion $X_{SF}$

Understandably, a CLC unit should convert the inserted fuel to completion to ensure an economic operation. If a fuel particle is transported out of the CLC reactor system before all combustible parts are converted, this fraction cannot be used to reduce the metal oxide OC. This effect can be mainly attributed to elutriation of char particles. The equation (12) correlates the gaseous carbonaceous flows out of the system to the carbon fed with the solid fuels. A value of 1 means that all carbon was converted to gaseous products, including combustible gases $CH_4$, $H_2$ and CO.

$$X_{SF} = \frac{\dot{n}_{C,gaseous,AR,out} + \dot{n}_{C,gaseous,FR,out}}{\dot{n}_{C,solid,FR,in}} \tag{12}$$

Oxygen demand $\Omega_{OD}$

If gaseous fuel is used or a solid fuel releases its volatiles and is gasified, mainly $CH_4$, $H_2$ and CO are produced. For economic reasons again, a complete conversion of those is desired. This is expressed by the oxygen demand, which correlates these gas flows of the combustible gases to the amount of oxygen needed for the entire injected fuel. A value of 0 indicates the complete gas conversion, which means an economic operation of the system.

$$\Omega_{OD} = \frac{0.5\dot{n}_{H_2,FR,out} + 0.5\dot{n}_{CO,FR,out} + 2\dot{n}_{CH_4,FR,out}}{\dot{n}_{O_2,\text{needed for stoichiometric combustion}}} \tag{13}$$

### 2.2.3. Pilot plant operation

In the general description of the CLC process it was reported that pilot plant operation with methane started 2001 by Lyngfelt et al. [6] at Chalmers University of Technology. It was described before that gaseous fuels are unsuitable for industrial scale units due to thermodynamic efficiency limitations. Therefore, the research community focuses mainly on the conversion of solid fuels nowadays. Two previous PhD theses at TUHH by Thon [15] and Kramp [16] covered the pilot plant operation up to the year 2012. In this chapter the advances in solid fuel conversion in CLC pilot plants starting from 2012 are summarized in Table 2.

**Table 2: Pilot plants operated with solid fuels since 2012 with the exception of work conducted at TUHH.**

| Group | Researcher | Year | fuel input(1) | fuel type | process route | OC | oxygen demand [%] | carbon capture [%] | fuel conversion [%] | fuel time [h] | ref |
|---|---|---|---|---|---|---|---|---|---|---|---|
| Chalmers University of Technology, Gothenburg, Sweden | Markström et al. | 2013 | 100 kW | bit. coal | iG-CLC | ilmenite | 15.9 – 21.6 | 96.4-99.5 | n.a. | > 5.6 | [48] |
| | Linderholm et. al | 2012 | 10 kW | bit. coal + pet coke | iG-CLC | ilmenite / Mn | 10 - 45 | 51 - 98 | 49 - 84 | ~ 40 | [49] |
| | Linderholm et al. | 2014 | 100 kW | wood char + bit. coals | iG-CLC | Tierga ore | 7.4 - 17 | < 94 | 71.4 | ~ 25 | [50] |
| | Schmitz et al. | 2016 | 100 kW | wood char + bit. coals | iG-CLC | ilmenite + Mn | 10-21 | > 98 | 0.55 – 0.77 | 18 | [51] |
| | Markström et al. | 2014 | 100 kW | pet coke + bit. coal | iG-CLC | Ilmenite | 14.8 – 25.7 | 74.9 – 96.0 | n.a. | 1.5 | [52] |
| | Schmitz et al. | 2016 | 10 kW | biochar | CLOU + iG-CLC | Ca + Mn | 2.1 – 9.0 | 77.4 – 95.2 | n.a. | 37 | [53] |
| Instituto de Carboquimica (ICB-CSIC), Zaragoza, Spain | Mendiara et al. | 2013 | 1.5 kW | pine sawdust | iG-CLC | Tierga ore | 3.8 – 14.1 | 97.5 | 78 - 87 | 37 | [54] |
| | Adanez-Rubio et al. | 2014 | 1.5 kW | pine wood chips | CLOU | Cu on Al/Mg | 0 – 18 (2) | 80 – 100 | 0.3 – 1.0 (3) | 10 | [55] |
| | Mendiara et al. | 2018 | 0.5 kW | waste biomass | iG-CLC | Tierga ore | 25 – 30 | 90 - 100 | 65 - 100 | 78 | [56] |
| | Perez-Vega et al. | 2016 | 50 kW | bit. coal | iG-CLC | ilmenite | 8.2 – 9.8 | 66.1 – 90.0 | 87.9 – 94.3 | n.a. | [57] |
| | Abad et al. | 2012 | 1.5 kW | bit. coal | CLOU | Cu on Al/Mg | 0 (2) | 96-99 | 93 – 97 | 18 | [58] |
| | Abad et al. | 2015 | 20 kW | bit. coal | iG-CLC | ilmenite | 7.4 – 8.6 (4) | 60.7 – 87.6 | 80.1 – 87.0 | n.a. | [59] |
| | Adanez-Rubio et al. | 2013 | 1.5 kW | bit. coal, lignite, anthracite | CLOU | Cu on Al/Mg | ~ 0 (2) | 75 – 99 | 75 – 99 (3) | 40 | [60] |
| Southeast University, Nanjing, China | Wang et al. | 2016 | 20 kW | Bit. coal | iG-CLC | iron ore | 7.23 | 97.1 | 92.7 | 98 | [61] |
| | Gu et al. | 2014 | 1 kW | Anthrac. coal | iG-CLC | Potassium-modified iron ore | n.a. | 93.7 | n.a. | 2 | [62] |
| Huazhong University of Technology, Wuhan, China | Ma et al. | 2015 | 5 kW | Bit. coal | iG-CLC | Hematite | 3.7 – 26.5 | 40 – 86 | n.a. | 300 (5) | [63] |
| Darmstadt University of Technology, Darmstadt, Germany | Ohlemüller et al. | 2016 | 1 MW | Hard coal | iG-CLC | Ilmenite | 22-28 | 44-54 | n.a | > 3.5 | [64] |
| | Ströhle et al. | 2014 | 1 MW | Hard coal | iG-CLC | Ilmenite | 12-17 | n.a. | n.a | ~6 | [65] |
| The Ohio State University, Columbus, USA | Bayham et al. | 2015 | 25 kW | Sub-bit. coal | iG-CLC | Iron oxide | 0.14 | 94,9 | ~100 | 70 | [66] |
| VTT, Espoo, Finland | Toni Pikkarainen et al. | 2016 | 50 kW | biomass (black + white pellets) | iG-CLC | ilmenite | 29 – 41 | 83 – 96 | n.a. | > 3 | [67] |

(1) Fuel input states the designed rated power of the plant. Usually the fuel input flow was lower for experimental reasons.

(2) Calculated from FR combustion efficiency, which is the opposite of the total oxygen demand $\Omega_T$. the total oxygen includes also unconverted solid fuels parts, i.e. the carbon fines. The regular oxygen demand $\Omega_{OD}$ is based only on unconverted fuel gases.

(3) Char conversion $X_c$, similar to fuel conversion $X_{SF}$, but only based on the char, which is actually converted in the FR.

(4) Total oxygen demand $\Omega_T$, which includes also unconverted solid fuels parts, i.e. the carbon fines.

(5) Time of hot fluidization, operation with fuel input lower

The circulating fluidized bed for energy purposes is a mature technology and, therefore, is used widely in the field of CLC. For the combustion of solid fuels, the research group at Chalmers University of Technology have operated two plants for solid fuel combustion of 10 kW and 100 kW, respectively, which run on ilmenite, Mn ores, and composites [48–53]. At Instituto de Carbochimica in Zaragoza three plants (0.5 kW, 1.5 kW, and 50 kW) for different solid fuels were constructed [54–60]. The plants were mostly run with oxygen carriers (OC) based on iron and copper oxide, however, waste materials from industrial processes were also investigated. The plant with 50 kW allows for chemical-looping combustion with oxygen uncoupling (CLOU) operation.

At Southeast University in China, a plant for the combustion of coal and biomass with a rated power of 1 kW was erected. Recently, the operation of a pressurized 20 kW system that runs on iron ore was demonstrated [61,62]. Hematite was used as an oxygen carrier to combust Chinese bituminous coal in a 5kW plant at Huazhong University of Technology [63]. The largest CLC plant for solid fuels with reported results is operated at TU Darmstadt with a rated power of 1 MW [64,65]. The research group at Ohio State University follows a different approach regarding the process layout. The FR is operated as a moving bed reactor, which showed good results in overall fuel conversion. At VTT a dual circulating fluidized bed was constructed, and both methane and biomass were successfully converted in that system [67].

Experiments with biomass as fuel have the advantage to result in negative emissions, so called Bio-Energy Carbon Capture and Storage (BECCS). Such experiments were carried out at Chalmers, CSIC and VTT between the years 2012 and 2018 [53,56,67]. As early as in 2009, the research group at Southeast University reported CLC operation with biomass [22].

Looking at the reported results in Table 2, one can see some trends regarding the used process route, the OC in the system and the fuel conversion behavior:

- The CLOU process route leads to advantageous conditions in the system, meaning that the elementary oxygen in the FR converts the fuel faster and to a higher degree than in iG-CLC. Very low oxygen demands are reported when using a CLOU-OC. Unfortunately, there is not yet a cheap material proposed as a CLOU-OC and only highly manufactured and expensive OCs were reported successful in the CLOU systems. Therefore, the search for cheap and abundant CLOU carriers, based on natural ores and waste materials is ongoing, which can be seen by the numerous publications covering mixtures of the materials.

- Most research was conducted with natural iron ores as an OC, especially ilmenite was widely investigated. This is due to the low cost compared to manufactured OCs, which could lead to economical operation. When a CLC unit is operated, attrition will constantly lead to the need for make-up material that needs to be added. The downside of the natural ores is the low reactivity with fuel gases, especially methane and the low oxygen carrying capacity per weight. This leads to a high amount of unconverted gases and, with it, the oxygen demand is high of mostly over 10%.

- Fuel composition has a major impact on the conversion in the system. Coals with a high amount of fixed carbon and little volatiles do usually induce a high carbon slip in iG-CLC systems. This lowers the carbon capture ability of the plants, because more char parts move to the AR where they react with air. This can be traced back to the slow gasification of the char in coals with high fixed carbon.

- The char structure of high volatile fuels, i.e. lignite and most biomass, was shown to be more reactive, as it is more porous and of lower density. These fuels showed low carbon slip and high carbon capture. For those fuels the amount of unconverted gases leaving the CLC system and, with it, the oxygen demand is significant.

The design choice for the iG-CLC plant at the TUHH is built on similar assumptions. For the efficient conversion of a fuel with high volatile content, a two-stage FR system was deployed. These two stages are used to reduce the amount of unconverted fuel gases tremendously.

## 2.3. Reaction kinetics

Reaction modeling

The results of a process simulation can be used for design, scaling and operation purposes. If a chemical reactor has to be modelled and simulated, one has several options to describe the events taking place in it. If one is not really interested in the detailed chemistry or the chemical reaction is assumed to be complete, one can simple use mass balancing to calculate a chemical reactor. This general mass balancing can be extended with knowledge about stoichiometric data, chemical yields or energy analysis to capture a process behavior, which is often done in commercial software like Aspen Plus.

Unfortunately, the chemical reactions, especially those involving solids, are influenced by numerous phenomena inside a reactor. These are the residence time of gases and particles, their respective concentrations and mixing behavior and certainly general operation conditions like temperature and pressure play a major role. The mixing behavior of gases and solids is often described by Continuously Stirred Tank Reactor (CSTR) or Plug Flow Reactor (PFR) models. The CSTR model assumes perfect mixing among the reactants in a reactor. The PFR model describes a one-dimensional movement of reactants through a geometry, without back and front mixing. With those two most extreme cases, the mixing behavior of most reactors can already be approximated accurately.

Knowing the whereabouts and the mixing behavior of gases, one can start to calculate the kinetic parameters for the prevailing operation conditions. For the simulation, a kinetic reaction constant k has to be derived from experiments, which at best should follow an Arrhenius type temperature behavior following equation (14):

$$k(T) = k_0 e^{-\frac{E_a}{RT}} \tag{14}$$

Arrhenius type kinetic parameters can be derived from a series of experiments at various temperatures and reactive gas inlet concentrations [68]. With the help of equation (15), one can determine the reaction order n, the activation energy $E_A$ and the pre-exponential factor $k_0$ for the reactions of the OC with the reactive gases. The char gasification is usually described with a set of Arrhenius reaction constants, which will be described later. The Arrhenius equation gives the reaction rate constant k, which has to be related to the prevailing gas concentration $C_g{}^n$ and the current solid conversion $X_s$ of the involved solid, OC or char. The reason for this is that the reactivity of the solid usually decreases with the progress of the reaction. Further, depending on the chosen reactor model, the reactive gas concentration $C_g$ is also declining, which has also a negative influence on the reaction rate at a certain time point. In equation (15) the solids conversion over time is described by the Arrhenius kinetic constant k(T) together with the gas concentration $C_g{}^n$ and the progress of the solid conversion $X_s$. In the equation, the term $f_{model}(X_s)$ is a structure factor and varies for different particle reaction models, which are described later. The kinetic constant k is the reciprocal value of the time for total conversion $\tau(T,\ C_g{}^n)$, which can be experimentally determined.

$$\frac{dX_s}{dt} = k(T) \cdot C_g{}^n \cdot f_{model}(X_s) \tag{15}$$

The equation needs the adaption of the units of the gas concentration $C_g$ according to the found reaction order n.

Particle reaction models

The reaction progress of OC material is usually investigated in kinetic reactors or with TGA methods. An OC sample is exposed to the reducing gases $CH_4$, CO and $H_2$, which are to be expected in a CLC environment. Afterwards the OC sample is oxidized with air or $O_2/N_2$ mixtures. This procedure is found in for example Garcia-Labbiano et al. [69] and Abad et al. [70], who used high volumetric flow rate TGA to overcome mass transfer limitations. The reduction and oxidation behavior are then compared to so called particle reaction models, which describe the progress of the reaction over time and/or conversion. Other research groups were trying to find a more complex reaction mechanism [71,72]. Kruggel-Emden et al. [73] conducted an overview over a plethora of used particle reaction models and found that most particle reaction models fail especially towards full conversion of OC particles.

In this work, the Homogeneous or Volumetric Reaction Model (VRM) and the Shrinking-Core Model (SCM) are investigated, whether the observed reaction pattern of the OC and char can be described accurately. More complex models were omitted as they usually do not give better results.

If the investigated particles react according to the VRM, a linear dependency of the rate of reaction with the remaining amount of reactive material is observed. The VRM supposes a reaction, which takes place over the whole volume of the particle [25]. Therefore, the solids conversion over time $\frac{dX_s}{dt}$ decreases linearly with the solids conversion $X_S$. The structure factor, hence, becomes (1 - $X_S$). For the progress of the conversion with time it becomes:

$$\frac{dX_s}{dt} = k_{VRM} \cdot (1 - X_S) \tag{16}$$

To describe the conversion state for a certain time of reaction, the equation (16) can be integrated to:

$$X_S = 1 - e^{k_{VRM} \cdot t} \tag{17}$$

If experimental results are to be evaluated, it can become beneficial to plot the model over the kinetic constant and the time. If the slope shows then a linear inclination, this means that the used particle reaction model is well suited to describe the experiment.

$$-\ln(1 - X_S) = kVRM \cdot t \qquad (18)$$

The SCM, as suggested by Kunii and Levenspiel [25], describes the solid reaction as taking place on the outer surface of an unreacted core. The core is shrinking during the solid reaction and, hence, the reactive surface as well. The structure factor for the SCM $f_{SCM}$ is dependent on phenomena that can take place, which influence the progress of the heterogeneous reaction. These are namely the film diffusion through the film layer around the particle, the ash diffusion through the ash deposit around the unreacted shrinking core and a general reaction limitation at the reactive surface. For the progress of the conversion over time the SCM with a chemical reaction limitation follows the equation (19).

$$\frac{dX_s}{dt} = k_{SCM} \cdot (1 - X_S)^{\frac{2}{3}} \qquad (19)$$

To describe the conversion state for a certain time t of reaction, the equation (19) can be transformed to:

$$X_S = 1 - \left(1 - \frac{1}{3} kSCM \cdot t\right)^3 \qquad (20)$$

The form to get a linear expression for the kinetic parameter $k_{SCM}$ is:

$$3[1 - (1 - X)^{\frac{1}{3}}] = kSCMt \qquad (21)$$

Figure 11 illustrates the effects of the different particle reaction models by plotting the conversion rate over the conversion for an arbitrary kinetic constant $k_{model} = 0.001$.

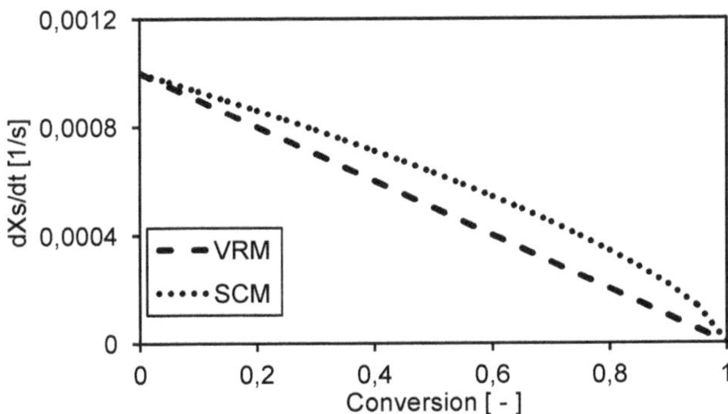


**Figure 11: Influence of the particle reaction model on the progress of the conversion rate for an arbitrary kinetic constant of kmodel = 0.001.**

Measurements techniques for reaction kinetics of heterogeneous reactions

Three measurement methods, which are used widely for investigating kinetics of reaction for fluidized bed applications, are depicted in Figure 12.

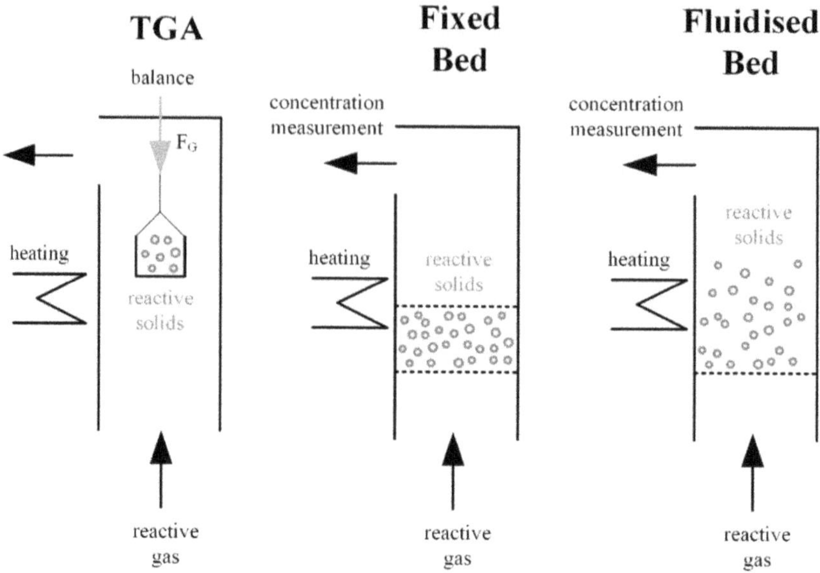


**Figure 12: Kinetic measurement reactors.**

The first one is the Thermogravimetric Analyzer (TGA) [74]. The basic principle is measuring the weight change of a solid sample, which is placed in an inert pan, usually made of platinum. The solid sample loses or gains weight according to the stoichiometry of the investigated reaction.

Another method is the exposition of solids, which are arranged in a packed bed, with a reactive gas. By analyzing the concentrations of the outflows and knowing the flow rates of the inflows, a molar balance can reveal the reactions taking place in the solids. There were some adaptions made to fixed bed reactors, to overcome some limitations, which lead to the Berty-reactor [75].

The last possibility, which is described in the Figure 12 is the laboratory scale fluidized bed kinetic reactor. In this reactor, the particle bed is fluidized with a reactive or inert gas. The reactive particles can be inserted with a pipe into an inert bed or be the starting bed material. Problems arise compared to TGA measurements because the gas concentration varies over the height of the fluidized bed reactor. Hence the reaction conditions along the reactor height are ever changing and make the direct derivation the kinetics parameters from measurements impossible. Therefore, gas and solids conversion should be calculated via a reactor model considering the hydrodynamics of the system, which was investigated at TUHH by Sitzmann [76]. Yu et al. [77] successfully overcame some of the limitations of a fluidized bed system in their micro-fluidized bed reactor (MFBR), to directly extract information about the conversion behavior. By keeping the size of fluidized bed sufficiently small, the complex flow structure of the fluidized bed can be kept negligible. All the proposed methods show certain advantages and disadvantages, which are summarized in Table 3, together with some basic information.

**Table 3: Measurement principle for kinetics of different reactor types. Approximated mixing behavior of solids and gases. Advantages and disadvantages of the reactors for kinetic measurements.**

| | TGA | Packed Bed | Fluidized Bed |
|---|---|---|---|
| measurement principle | gravimetric | gas concentration (molar balance) | gas concentration (molar balance) |
| Mixing behavior gases | CSTR | PFR or CSTR via recycle gas | PFR |
| Mixing behavior solids | CSTR | PFR | CSTR |
| advantages | + replicability<br>+ small sample size<br>+ no time delay of weight measurement<br>+ constant gas concentration | + no meso flow structures<br>+ constant gas concentration | + mixing behavior<br>+ flow structures as in real fluidized bed<br>+ same phenomena<br>+ heating rate |
| disadvantages | - heating rate<br>- diffusion effects (film layer) | - mixing<br>- replicability<br>- time delay of gas measurement | - replicability<br>- time delay of gas measurement<br>- concentration gradient over height |

Reaction kinetics of the OC

In flowsheet simulation a complete process is modeled and, hence, each single process unit and all phenomena in these units should be described as simple as possible to save computing time. This means for the fluidized bed reactors that the conversion of chemical compounds is rather represented via overall reactions than by a detailed reactions scheme with intermediate products. In this work, mainly an OC based on copper oxide as the reactive metal oxide is used. Coal devolatilization and gasification is expected to yield the combustible gases $CH_4$, CO and $H_2$. For CLC conditions, Abad et al. [78] found that these gases are transformed by a CuO-OC as the active material via the equations (23)-(25).

$$CH_4 + CuO -> CO + 2H_2 + Cu \quad (22)$$

$$CO + CuO -> CO2 + Cu \quad (23)$$

$$H_2 + CuO -> H_2O + Cu \quad (24)$$

The active compound on the carrier is re-oxidized via:

$$Cu + ½ O_2 -> CuO \quad (25)$$

This implies that the methane is partly oxidized first to CO and $H_2$ and these intermediates react further to $CO_2$ and $H_2O$. Copper oxide carriers are reported to be able to fully convert methane towards the products $CO_2$ and $H_2O$ and hence one can also think of the direct expression in equation (26).

$$4\,CuO + CH_4 -> 4\,Cu + CO_2 + 2H_2O \quad (26)$$

Adanez et al. [10] summarized all reaction kinetic analyses conducted in the field of CLC until 2012 and represents the basis of Table 4. There is a mere amount of kinetic studies on other OC materials described there, but only the ones focusing on a CuO-based carrier or studies conducted in a laboratory scale fluidized bed are included in the Table. This is since a

CuO-based carrier was used throughout the work presented here and a laboratory scale fluidized bed reactor was operated for kinetic studies. Work by Lindmüller [79], which was done in conjunction with this PhD project, added recent studies with CuO-based OC carried out in fluidized bed reactors.

**Table 4: Reaction kinetic studies with CuO oxygen carriers and studies using a fluidized bed laboratory reactor.**

| Oxygen-Currie | | Experimental conditions | Kinetic data | Ref. |
|---|---|---|---|---|
| 82 wt% CuO on $Al_2O_3$<br>ROC = 16% | $D_p$ = 355 – 500 μm<br>ε = 75% | Fluid. bed T = 250-900 °C<br>2-10 vol% $H_2$<br>($2CuO \rightarrow Cu_2O$)<br>2-10 vol% $H_2$<br>($Cu_2O \rightarrow Cu$)<br>2-10 vol% CO<br>($2CuO \rightarrow Cu_2O$)<br>2-10 vol% CO<br>($Cu_2O \rightarrow Cu$) | SCM, n = 1.0<br>$E_A$ = 58 kJ/mol<br><br>$E_A$ = 44 kJ/mol<br><br>$E_A$ < 80 kJ/mol<br><br>$E_A$ = 56 kJ/mol | Chuang et al.[72,80,81] |
| 60 wt% CuO on $Al_2O_3$(23%) CaO (17%), $R_{OC}$ = n.a. | $D_p$ = 300- 425 μm<br>ε = 36% | Fluid. bed T = 850-950 °C<br>5% $H_2$ | VRM, n = 1.0<br>$E_A$ = 59.7 kJ/mol | Hu et al. [71] |
| 99 wt% $Fe_2O_3$,<br>$R_{OC}$ = 3.3% | $D_p$ = 300 – 425 μm<br>ε = 60% | Fluid. bed T = 250-900 °C<br>1 -9 vol% CO | SCM, n = 1.0<br>EA = 75 kJ/mol | Bohn et al.[82] |
| 14 wt% CuO on $Al_2O_3$,<br>$R_{OC}$ = 2.8% | $D_p$ = 100-500 mm<br>ε = 53% | TGA T = 600-800 °C<br>5-70 vol% $CH_4$<br>5-70 vol% $H_2$<br>5-70 vol% CO | SCMg with Fg = 1<br>n = 0.5 $E_A$ = 106 kJ/mol<br>n = 0.5 $E_A$ = 20 kJ/mol<br>n = 0.8 $E_A$ = 11 kJ/mol | Abad et al. [78] |
| 10 wt% CuO on $Al_2O_3$,<br>$R_{OC}$ = 2.0% | $D_p$ = 100-300 mm<br>ε = 57% | TGA T = 600-800 °C<br>5-70 vol% $CH_4$<br>5-70 vol% $H_2$<br>5-70 vol% CO | SCMg with Fg = 1<br>n = 0.4 $E_A$ = 60 kJ/mol<br>n = 0.6 $E_A$ = 33 kJ/mol<br>n = 0.8 $E_A$ = 40 kJ/mol | García-Labiano et al. [69] |
| 60 wt% $Fe_2O_3$–20 wt% CuO on $Al_2O_3$, $R_{OC}$ = n.a. | $D_p$ = 106- 250 μm<br>ε = n.a. | TGA, T = 600-950<br>3 vol% H2 | VRM, n = 1.0<br>EA = 42.6 kJ/mol | Ksepko et al. [83] |
| 35.9 wt% CuO on $SiO_2$,<br>$R_{OC}$ = n.a. | $D_p$ = 50 – 150 μm<br>ε = n.a. | TGA, T = 800-850°C<br>10% H2 (2CuO→Cu2O)<br>10% H2 (Cu2O→Cu)<br>5% CO (2CuO→Cu2O)<br>5% CO (Cu2O→Cu) | AEM (1),<br>ν= 1.1, n = 1.0<br>$E_A$ = 65.1 kJ/mol<br>$E_A$ = 69.8 kJ/mol<br>$E_A$ = 101 kJ/mol<br>$E_A$ = 127 kJ/mol | Zhou et al. [84] |

(1) AEM = Avrami-Erofeev Model, a mathematical fitting model.

Looking at the different kinetic studies in the Table 4, one can see that no clear trend for the reaction order or the activation energy $E_A$ is visible for CuO-based OCs. Even within one research group, different parameters can be found for similar carriers [69,78]. This can be explained with the complex structure of these solid materials. First, the preparation method has a major influence on the pore structure and availability of the reactive compound. Further, the reduction and oxidation cycles in a fluidized bed have also a major influence on the solid structure [85], due to attrition, thermal stress and chemical conversion. One major change occurring with CuO-based OCs in concourse with $Al_2O_3$ base is the formation of a copper aluminate spinel $CuAl_2O_4$. This structure also has shown to exhibit oxygen carrying capabilities similar to CuO [86]. The main difference lies in the ability of releasing elementary oxygen, which is lost if a spinel is formed. It was shown that after 50 redox cycles, the entirety of the Cu has formed the spinel structure, when $Al_2O_3$ was the base material. This could be confirmed in own research, when the bed material of the pilot plant operated for over 100 h was investigated via X-ray diffraction.

Reaction kinetics of fuels under CLC conditions.

Lignite gasification in the fluidized bed dates to 1922, when a fluidized bed reactor was first applied to generate syngas for ammonia fertilizer production by Winkler [87] at IG Farbindustrie, nowadays BASF. In the last decades, the reaction kinetics of char gasification in inert bed fluidized bed reactors was widely investigated and reference literature, e.g. Johnson et al. [88], describe the mechanisms of $CO_2$-char gasification, steam char gasification and pressurized gasification. The basic gasification reaction equations (27) and (28) describe the conversion of the main reactants in this heterogeneous gas-solid reaction. The complexity of steam and $CO_2$ gasification is increased by the homogeneous WGS reaction, eq.(29), because the gasification agents $H_2O$ and $CO_2$ are converted into each other depending on the equilibrium of the reaction.

$$C_{char}(s) + CO_2(g) \leftrightarrow 2\,CO(g) \tag{27}$$

$$C_{char}(s) + H_2O(g) \leftrightarrow H_2(g) + CO(g) \tag{28}$$

$$CO(g) + H_2O\,(g) \leftrightarrow CO_2(g) + H_2(g) \tag{29}$$

According to Matsui et al. [89–91], Langmuir Hinshelwood type kinetics can be applied for both steam- and $CO_2$-gasification of coal char, which are defined in equation (30) for steam and (31) for $CO_2$.

$$\frac{dX_s}{dt} = \frac{k_1[H_2O]}{1+k_2[H_2O]+k_3[H_2]+k_4[CO]}\,f\,(X_s) \tag{30}$$

$$\frac{dX_s}{dt} = \frac{k_1[CO_2]}{1+k_2[CO_2]+k_3[CO]}\,f\,(X_s) \tag{31}$$

In equation (29(30), $k_1[H_2O]$ represents the forward reaction of gasification via steam, $k_2[H_2O]$ the backward reaction, $k_3[H2]$ and $k_4[CO]$ the inhibiting effects of hydrogen and carbon monoxide on steam gasification. In equation (31), the gasification agent is $CO_2$ and $k_1[CO_2]$ represents the forward reaction of gasification via $CO_2$, $k_2[CO_2]$ the backward reaction and $k_3[CO]$ the inhibition by carbon monoxide. The brackets around the gases indicate a concentration dependence of the kinetic parameters. One can see that the solids conversion per unit time $\frac{dX_s}{dt}$ is on the one hand depending on the kinetic constants and the kinetics of each participating gas species and on the other hand depending on the state of the char conversion $X_s$. This latter phenomenon is expressed by the function $f(X_s)$, which again can take different forms according to the particle reaction mechanism, which was described in this section before.

In the FR of an iG-CLC plant, the solid fuel is at first converted to syngas, which then can react further with the OC. This means that in contrast to the gasification in an inert sand bed, the main gasification products CO and $H_2$ are constantly removed from the gas stream in the reactor. The effect of CO and $H_2$ on the gasification rates is already included in the mechanisms above as they appear in the denominator of eq. (4) and (5). The removal of those inhibiting gases could lead to a major enhancement of the gasification rate and the process performance of CLC. In a fluidized bed reactor, the OC conversion products $CO_2$ and $H_2O$ will also contact again with unreacted coal char, which will also enhance gasification.

The heterogeneous reaction between the fuel gases $CH_4$, CO and $H_2$ with the OC can be investigated via thermogravimetric analyzing (TGA) methods. However, the kinetics of the char conversion in a reactive fluidized bed cannot sufficiently be determined in a TGA. This is due to gas-solid mixing and fluid mechanics conditions inside a fluidized bed, which are impossible to achieve in a TGA pan. Hence the interactions of the two reactive solids and their implications should if possible be investigated in a fluidized bed reactor. The conversion kinetics of solid fuels is a major design criterion for iG-CLC as the slip of unconverted char from the FR to the AR has to be avoided. To investigate the effects of reactive bed material, like ash or the OC, on the conversion, fluidized beds must be used. Unfortunately, with the application of fluidized beds, problems arise in the evaluation of the experimental data. First, the effects of the char conversion cannot be directly and instantaneously measured as in a TGA but have to be derived from mass balances established via gas concentration measurements. Additionally, hydrodynamic phenomena, like bubbles in a bubbling fluidized bed, lead to a bypass of gas, which does not take part in the heterogeneous reactions. This will cause concentration gradients between unreactive bubbles and the reactive suspension of solids and gas and, with it, diffusive and convective mass transfer will take place. When considerable amounts of gas are converted also gas concentration gradients over the height of the reactor can occur and hence the kinetic assumption of a certain operation condition, e.g. gasification in 25 vol-% $CO_2$, is not true. A Possible solution to overcome these difficulties is the application of a detailed model to interpret the effects of fluid mechanics and gas gradients. Another solution would be the usage of a micro fluidized bed reactor [77], which shows low gradients and minimal effects of large scale fluid mechanics.

Troiano et al. [92] introduced lignite char of the particle size ranging from 2 mm to 6 mm into a fluidized bed reactor with sand as bed material in separate $H_2O$ and $CO_2$ atmospheres. They state that the gas generation due to the gasification was low compared to the overall gas flows. This means that the hydrodynamic effects of the fluidized bed could be disregarded, and a reactor model was not applied. Cuadrat et al. [93] conducted fluidized bed experiments with Norwegian ilmenite (150 - 300 µm) as OC and South African bituminous coal [100 - 200 µm]. They found an increase in the instantaneous rate of gasification of around 10 - 70% over the course of conversion, when ilmenite was used as a bed material compared to inert bed experiments. For the Columbian coal "El Cerrejon", the same group found a set of kinetic parameters for gasification, assuming homogeneous reaction and chemical reaction limitation of the process [94]. Scott et al. [95] tested a copper oxide OC in a fluidized bed reactor and conducted gasification of lignite in a bed of the carrier. In a next step the same group (Brown et al. [96]) investigated the gasification of lignite char in a fluidized bed reactor. The bed material was first inert sand and was then exchanged with an iron oxide OC. The kinetic rate constants for a Langmuir-Hinshelwood approach could be extracted. Both batch experiments as well as continuous coal feeding was experimentally conducted. A reactor model was used to simulate the carbon holdup in the reactor. An increase in the rate of gasification was traced back to the conversion of inhibiting CO to $CO_2$. Keller et al. [97] used sand and two oxygen carriers, ilmenite and oxide scales from steel production, to investigate the inhibition of the gasification reaction by hydrogen. In their setup the gasification was not influenced by any CO concentration inserted with the gasification agent. Inhibition effects could be observed for hydrogen and modeling of the inhibition was successful. Arjmand et al. [98] tested 8 different manganese materials and ilmenite for their conversion behavior of gases and two solid fuel chars. They could prove that the manganese materials accelerate the gasification to a higher degree than ilmenite does. The rate change was attributed mainly to the better conversion of CO and $H_2$, because the reactivity of manganese was higher towards CO and $H_2$ than measured for ilmenite.

## 2.4. Modeling of Chemical Looping Combustion

Process simulation of CLC has been carried out with a variety of methods and approaches. In literature, no sharp separation of process simulation tools exists, but one can make some basic classification among the time and length scales, which the simulations have considered. One can, for example, distinguish dynamic and steady state simulations. In dynamic or also called transient simulations, the operation parameters and, hence, the fluid mechanics in the reactors can vary over time, which is closer to reality. Steady state simulations usually find one "steady state" solution for a set of operation conditions and is, therefore, less computationally expensive. Different simulation environments and ideas were used to model the whole process of CLC. Table 5 gives a short overview of flowsheeting software used for CLC process simulation by various research groups.

**Table 5: List of process modeling approaches by various research groups, (iG = in-situ Gasification).**

| Process | Fuel | Software | Fluid mechanics | Reaction model | Institution | Researcher | Dynamic |
|---|---|---|---|---|---|---|---|
| CLC | gas | IPSEpro | 0-D | equilibrium reactions | Vienna University of Technology | Bolhar-Nordenkampf et al. [99] | no |
| CLC | gas | Matlab/ Simulink | 1-D | reaction kinetics | Lappeenranta University of Technology | Peltola et al. [100,101] | (yes) (1) |
| CLC | gas | AspenPlus | 1-D | reaction kinetics | Heriot-Watt University Edinburgh | Porrazzo et al. [102] | no |
| Syngas-CLC | syngas | AspenPlus | 0-D | equilibrium reactions | University of Surrey | Mukherjee et al. [103] | no |
| Syngas-CLC | syngas | AspenPlus | 0-D | equilibrium reactions | Ohio State University | Li et al. [104] | no |
| iG-CLC | solid | AspenPlus | 0-D | equilibrium reactions | University of Utah | Sahir et al. [105] | no |
| iG-CLC | solid | AspenPlus | 1-D | reaction kinetics | Technical University of Darmstadt | Ohlemüller et al. [106] | no |
| iG-CLC | solid | SolidSim | 1-D | reaction kinetics | Hamburg University of Technology | Kramp et al. [16,107] | no |

(1) The work of Peltola was based on dynamic equations but was used to calculate steady state operation only.

Process simulations also vary in the depth of phenomena they consider. When looking at smallest structures of fluid mechanics, like particle-gas interaction in CFD or detailed chemical reaction schemes, the simulation is carried out on the micro or even molecular scale. When some phenomena are averaged, or closures are used to describe the micro scale behavior, one can speak of meso-scale simulations. In macro-scale simulations, only global mechanisms are resolved. This means that for certain unit operations, mass and energy balances are conducted, whereas in fluidized bed reactors also simplistic models for describing the solids distribution in the reactors can be used. For the different simulation approaches it can be said that the more phenomena included, the more computationally expensive the simulation will be. On the other hand, resolving the micro-scale phenomena accurately gives obviously more comprehensive results.

Flowsheeting is a process simulation tool, which considers the predominant macroscopic effects in a complex process network involving multiple process units. For describing these effects, semi-empirical and empirical models are used to characterize the events in each process unit to keep computational time and effort within manageable limits.

Two main approaches of solving the model equations in a flowsheet simulation environment exist. The first is the equation-oriented approach, in which all equations of the process units are concluded and solved simultaneously. The other is the modular-sequential approach, in which the process units, so called modules, are calculated in series. The big advantage of flowsheeting compared to CFD methods is the short computation time, which makes simulations over a long period of time possible (>1000s), provided that a dynamic approach is used. With this method, effects over a longer time can be tracked, like attrition, start-up and shut-down procedures, and load changes, which cannot be resolved within CFD.

The used approaches deviate considerably in the detail in which certain phenomena were resolved. 1-D models consider basic fluid mechanics of solids and gases inside the fluidized bed reactors. Usually one can find a dense zone at the bottom of the reactor, followed by a steady decline of solids concentration with increasing height above the distributor. In two-phase models, the dense zone is further divided into a bubble and a suspension phase. 0-D models do not resolve any of these basic phenomena commonly seen in fluidized bed reactors but try to mimic the outcome via a series of plug flow reactors (PFR) or continuously stirred tank reactors (CSTR). Recently, in AspenPlus a simple fluidized bed reactor module was implemented, but it was concluded that the model did not describe the hydrodynamics of a circulating fluidized bed (CFB) boiler sufficiently [108].

From the chemical reaction point of view, most researchers from Table 1 used a chemical equilibrium model, which minimizes the Gibbs free energy, without taking into account stoichiometry and reaction rates. Two groups used their own code in Matlab/Simulink and SolidSim, respectively, to implement gasification and char combustion kinetics [16,101]. Another group developed a fluidized bed reactor model and covered these latter reaction schemes by implementing their model in AspenPlus [106].

Dynamic modeling of CFB systems was rather seldom conducted. One first work was published recently by Panday et al. [109], who made a series of experiments on their 16 m high CFB system. Afterwards it was checked if empirical and semi-empirical models can capture the hydrodynamics. Further, their system was investigated towards the system's reaction on step changes. Hartge et al. [110] were looking at the dynamic effects of Particle Size Distributions (PSD) on the CFB behavior. All proposed process models concluded in Table 5 applied to CLC contain no time dependency, so they yield steady state results. The work of Peltola et al. [101] was including time dependent variables, but the system itself was used to model steady state operation. Apart from these simulations covering the whole process, a lot of attention was devoted to fuel reactor (FR) modeling, as the FR is crucial for the conversion of the solid fuel and, hence, the overall efficiency of capturing carbon dioxide. Since the FR is the crucial part of the process, many researchers have focused solely on the local fluid mechanics and reaction mechanisms in the FR [111,112].

The performance of the chemical looping process is predominantly decided by the conversion of combustible gases and solids in the fuel reactor. High reactivity of the OC can prevent gases from leaving the reactor unconverted. Further, if the char is not gasified sufficiently, it will be drawn over to the air reactor of the system and that will lower the carbon capture efficiency of the process. Therefore, many researchers put a special focus on the fuel reactor modeling. As Abad et al. [23] state in their publication on fuel reactor modeling, the detail of the modeling itself can vary and includes simple models without fluid mechanics, 1-dimensional fluid mechanic models based on empirical or semi-empirical models and 3-dimensional CFD models. Porrazzo et al. [102]compared CFD results with ASPEN Plus simulations using Plug Flow and Continously Stirred Tank reactor modeling to mimic fluid mechanics in the fuel reactor. In both settings, gasification kinetics from Cuadrat et al. [93,94] mentioned before were used. Ohlemüller et al. [106] used ASPEN Plus and implemented a 1-D fluid mechanic model with ASPEN Custom Modeler, for a CLC process simulation. There, also the kinetics from Cuadrat et al. were used. Wang et al. [61] conducted multi-scale CFD simulation including mass and heat transfer as well as chemical reactions with kinetics from Abad et al. [78], but only considered gaseous fuels. Modeling of CLC in the present author's group, started by Kramp et al. [16,107] for gaseous fuels, was conducted with SolidSim software using 1-D fluid mechanic modeling and gas-phase kinetics by Abad et al. [113] from thermogravimetric analysis (TGA). Summarizing the process modeling work on CLC, one can see that the used ch0ar gasification kinetics are based only on very few experiments, often derived from TGA measurements. The TGA is not able to reproduce the mass transfer phenomena between the gas phase and the solids, which can be seen in a fluidized bed.

For a more realistic fuel reactor modeling the kinetics of the conversion of the OC by the fuel gases is needed to estimate the conversion of gases, which is a key criterion for the performance of the whole process. In addition to that, the reaction behavior of the solid fuel will determine how much fuel will be carried over to the air reactor and hence lower the carbon capture. Unconverted fuel particles can also leave the reactor system with the fuel reactor off-gas, which decreases the FR combustion efficiency.

# 3. Methodology

In this work three different research tasks for CLC were investigated. The main task was the process simulation based on dynamic flowsheeting. To validate the simulations and the creation of models, a pilot plant for solids fuels was used with gaseous and solid fuels. For parameterization of the kinetic models in the process simulations, a laboratory scale reactor was run with different fuels, gases and at a variety of process conditions. The experimental findings are described first to better understand the general CLC process and the used plant.

## 3.1. Experimental pilot plant

At TUHH an experimental facility with a rated fuel input power of 25 $kW_{th}$ is operated. To tackle the previously mentioned problem of unconverted gas the plant was designed with two stages for the FR. The two stages are fluidized beds operating at bubbling operation conditions. The reactor was planned and constructed by the two previous PhD students Kramp [16] and Thon [15].

### 3.1.1. Layout and dimensions of the CLC pilot plant

In the first campaigns the fuel was injected into the lower of the two stages for gasification and devolatilization. The produced gases rise and are mixed into a bed of freshly oxidized OC from the AR. A photo of the 25 $kW_{th}$ plant is shown in Figure 13 and the basic layout of the plant is presented in Figure 14. Additionally, the movement of the circulated OC is depicted with the arrows. Plant dimensions are provided in Table 6. The entire reactors are surrounded by an electric heating jacket, which can heat the system up to a temperature of 1000°C. This guarantees a uniform temperature in all system parts. Temperature is controlled by a series of programmable logic controllers, which adjust the heating of ten electrical heating shells distributed over all process equipment involved. Keeping AR and FR inside one electrical oven means that both reactors are always kept at similar temperatures, which may be different on the industrial scale.

**Table 6: Dimensions of the CLC system operated at TUHH.**

| Component | Height [m] | Diameter [m] |
|---|---|---|
| air reactor (AR) | 8 | 0.1 |
| fuel reactor (FR) | 4 | 0.25 |
| cyclone | 0.34 | 0.21 |
| | **Width [m]** | **Length [m]** |
| loop seal 1 (S1) | 0.16 | 0.13 |
| loop seal 2 (S2) | 0.13 | 0.25 |

**Figure 13: Photo showing the insulated top of the pilot plant by Thon et al. [114].**

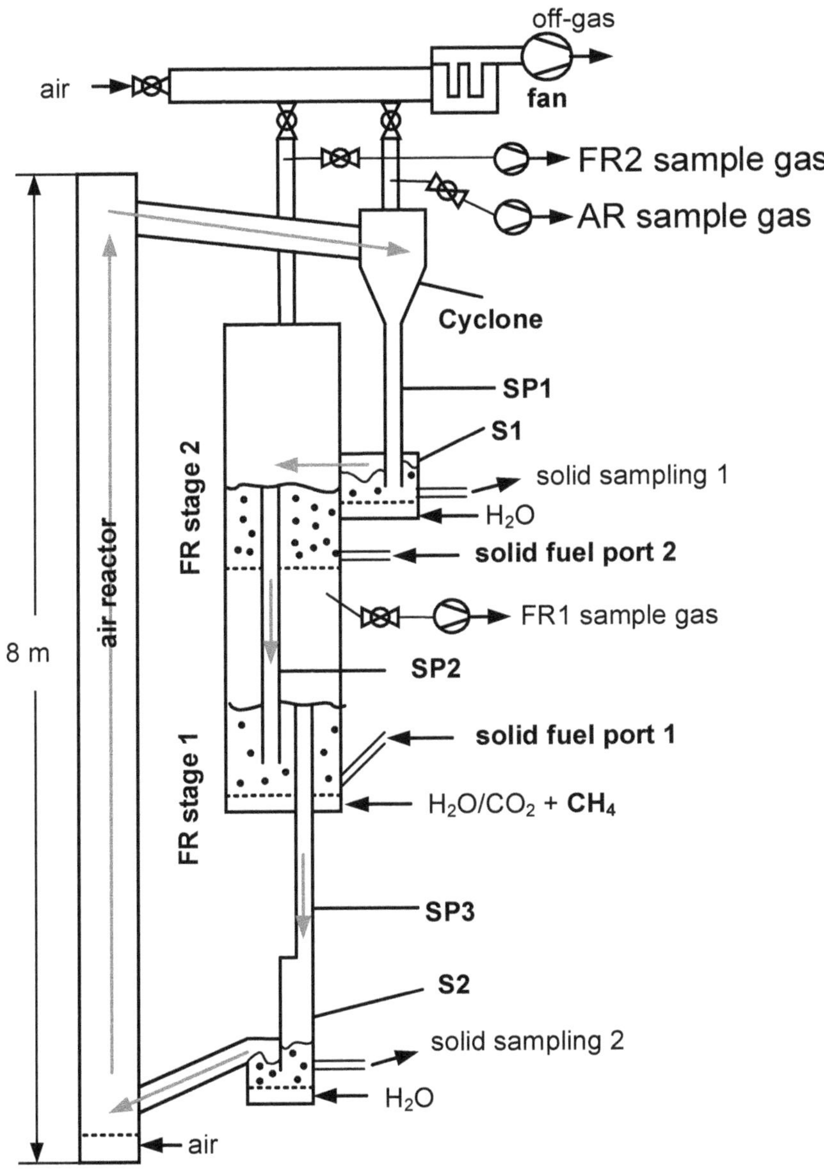


**Figure 14: CLC pilot plant including the different fuel injection ports and sampling stages. The grey arrows show the circulation direction of the solid OC in the plant.**

The air reactor is a CFB riser reactor. It is equipped with pressure sensors over the entire height to measure the pressure drop. The fluidizing agent is ambient air introduced at gas velocities of 3 - 6 m/s. The oxygen carrier is oxidized due to the contact with air at high temperatures. The regenerated material then passes into a cyclone, which is used for the separation of particles from the exhaust gas. After that the solids fall into the upper loop seal S1. It is installed for avoiding gas exchange between the reactors. The carrier material then enters the upper stage of the fuel reactor at fully oxidized state. The solid fuel, which is pneumatically fed by a lance, can be added in the upper stage as well as in the lower stage. The two stages can be fluidized with $CO_2$, compressed air and steam, usually at gas velocities under hot conditions of 0.15 - 0.35 m/s. Depending on the type of fuel, the supply location can be adjusted and thus the conversion efficiency can be influenced. The system circuit closes via the second loop seal, which leads the reduced carrier back into the air reactor.

To get an impression of the measuring system, a flow chart is shown in Figure 15. The exhaust gas analysis of both stages of the FR1 and FR2 can be carried out separately, which provides accurate results on fuel conversion and gasification states in both stages. The dust analysis is carried out by the evaluation of the dust trapped in the off-gas filters.

Outlet concentrations from the FR, are measured by nondispersive infrared sensors (NDIR) in the ranges of 0 to 25 vol.% of CO and $CH_4$ and 0 to 100 vol.% of $CO_2$. Hydrogen is measured via thermal conductivity in the range of 0 to 25 vol.% and compensating for cross interference with other gases. This is carried out in a 4-channel measurement equipment (fabricated by TAD GmbH in Unna, Germany, type GME.84-K4). Oxygen measurement from the FR outlet is done in a paramagnetic cell (Magnos 3K by Hartmann and Braun/now ABB). Off-gas samples were taken into gas bags for later analysis via gas chromatography. This was done to cross check gas measurements, because $CO_2$ concentrations close to 100% are reached inside the FR. The $CO_2$ and $O_2$ concentrations from the AR outlet are detected by electrochemical and paramagnetic devices ($O_2$-el.chem. by TAD, type GME.42.CD-O2.ec.d; $CO_2$ by NDIR sensor of Edinburgh Instruments). Concentration measurements can be conducted from both FR stages, to examine the effect of the second stage separately. Pressure measurement ports are employed to provide a precise picture of the pressure situation inside and between the reactors.

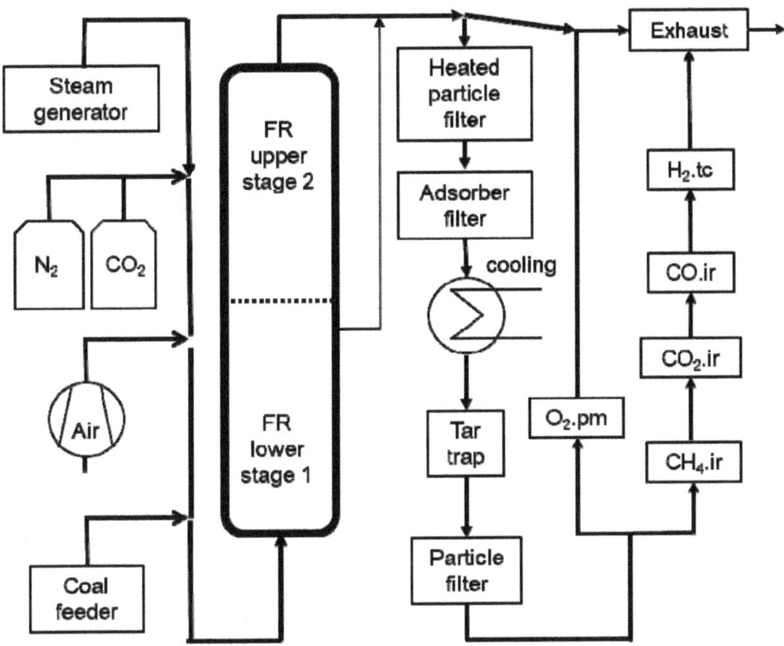


**Figure 15: Fuel reactor system with the gas cleaning and measurement lines.**

### 3.1.2. Materials used inside the CLC pilot plant

The solid fuels used in the experimental tests were bituminous coal, Rhenish lignite and German hard wood biomass. The hard wood biomass was produced by grinding German hard wood pellets. The Rhenish lignite was provided by RWE Germany, the bituminous coal originates from the USA and was supplied by Vattenfall. The chemical composition of the fuels is summarized in Table 7. To investigate the influence of the particle size certain size fractions of the solid fuels were sieved and separated. The cumulative distributions of those are shown in Figure 16. Fuels were injected in both fuel ports 1 and 2 from Figure 14

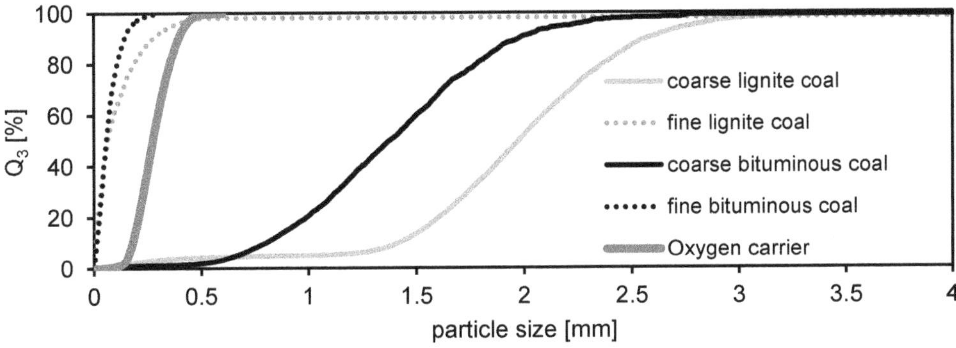


**Figure 16: Cumulative size distribution $Q_3$ of all materials in the CLC pilot plant.**

**Table 7: Solid fuel properties, proximate and ultimate analysis.**

| Property | Lignite (coarse/fine) | Bituminous coal (coarse/fine) | German wood biomass |
|---|---|---|---|
| Bulk density [kg/m3] | 1000 | 1400 | 354 |
| Lower heating value [MJ/$kg_{raw}$] | 22.2 | 26.7 | 18.7 |
| **Proximate Analysis** | | | |
| Ash [wt.% raw] | 4 | 12.9 | 0.2 |
| Moisture [wt.% raw] | 11 | 3.5 | 8.6 |
| Volatiles [wt.% raw] | 45 | 30.7 | 85.5 |
| Fixed carbon [wt.% raw] | 40 | 52.9 | 5.7 |
| **Ultimate Analysis** | | | |
| Carbon [wt.% raw] | 59.5 | 66.9 | 47.1 |
| Hydrogen [wt.% raw] | 4.3 | 4.5 | 10.3 |
| Oxygen [wt.% raw] | 20.3 | 10.1 | 31.9 |
| Nitrogen [wt.% raw] | 0.7 | 1.1 | 2.9 |
| Sulphur [wt.% raw] | 0.35 | 1 | < 0.1 |

A synthetic copper-based metal oxide sprayed on alumina support was used during all experiments in the pilot plant as an OC. The OC was used inside the plants for several experimental campaigns and added up for about 80 kg of bed material. Sometimes newly produced $CuO/Al_2O_3$ was added to make up for OC losses due to abrasion. Nevertheless, the median hot operation time of OC inside the plant is well over 100h. The properties are summarized in Table 8. The oxygen carrying capacity $R_O$ was experimentally determined by the weight difference of a sample, which was fully oxidized by air and later entirely reduced by $H_2$. Compared to the initial value of around $R_O$ = 2.5 %, the value of the bed material fell to 1.5 % due to abrasion of the outer layer of copper oxide from the support alumina.

**Table 8: Basic properties of the used oxygen carrier material.**

| Parameter | Oxygen carrier |
|---|---|
| Molecular composition | |
| Density [kg/m$^3$] | 4094 |
| Bulk density [kg/m$^3$] | 1027 |
| $R_0$ $[wt.\%]$ | 1.50 |

### 3.1.3. Operation of the CLC pilot plant

The CLC pilot plant was operated under hot fluidization conditions on over 50 experimental campaign days. There were several challenges during start-up and connected to measurements, which caused numerous shut-downs and maintenance stops. The main problems arose due to the clogging of the solid fuel feeding system, polluted measurement devices, faulty gas concentration measurements due to leakage into the measurement devices and problems with circulation and fluidization stability in the system. Overcoming these obstacles and optimal operation was described in an overhauled operation manual. Nevertheless, 41 successful and different operation conditions could be established in the reactor system with fuel feeding to the reactor. The operation with fuel feeding to the reactor was considered successful if the following criteria were met:

- homogeneous temperature profile in both reactors within ±10°C.
- stable gas concentrations in both FR and AR over at least 10 minutes.
- constant fuel mass flow from the balance of the fuel bunker and from AR off-gas concentrations.
- leakage of air ($N_2$ and $O_2$) to the FR gas below 15 vol.%.

A steady state operation of the system was usually established within 15-20 minutes, which is a relatively long time compared to regular air-fired combustors. This is due to the reactive oxygen carrier, which constitutes a big source of oxygen, when fuel operation is started. The fuel operation time could in most experiments be kept far above 30 minutes. Some shorter experimental time frames were considered enough, if the mass balancing from the experimental data showed clear results. The leakage could be arranged in most experiments below 5 vol.% of combined $N_2$ and $O_2$. When low fuel mass flows were used, e.g. with biomass or methane, the leakage mass flow of $N_2$ and $O_2$ was higher, which was traced back to both a low volumetric flow rate of apparent reactor gases as well as unfavorable pressure conditions in the FR. Previous research investigated the solid fuel conversion of lignite dust with ilmenite and the $CuO/Al_2O_3$ OC [114]. In Table 9 all experimental campaigns, which were conducted for this work are specified. The experimental campaigns shown here were carried out for following research targets:

- #1 - #6: Conversion behavior of fine bituminous coal.
- #7 - #9: Conversion behavior of coarse bituminous coal.
- #10 - #13: Conversion behavior of lignite dust with high fuel flows.
- #14 - #15: Establishing similar fuel flows of fine and coarse bituminous coal.
- #16: Injecting fine bituminous coal into the upper stage for better char conversion.
- #17 - #19: Injecting coarse bituminous coal into upper stage for better char conversion.
- #20 - #22: More experimental data on upper stage injection of fine bituminous coal. Load changes applied to investigate dynamic behavior.
- #23 - #25: Experimental data for upper stage injection of lignite dust.
- #26 - #30: More experimental data on upper stage injection of coarse bituminous coal.
- #31 - #35: Conversion behavior of coarse lignite. Injection in lower stage.
- #36: Dynamic operation with lignite dust and solid sampling
- #37: Conversion behavior of methane injection with the fluidization gas. Dynamic operation with solid sampling (fuel injection start/fuel stop).
- #38: Dynamic operation with lignite dust and solid sampling (fuel injection start/fuel stop).
- #39: Conversion behavior of coarse wood biomass.
- #40 - #41: Conversion behavior of fine wood biomass. Dynamic experimental results were produced as well (fuel injection start/fuel stop).

**Table 9: Investigated fuels and operation conditions in the CLC pilot plant.**

| # | fuel type | size fraction | T [°C] | mass flow [kg/h] | fuel power [$kW_{th}$] | fuel operation (1) [min] | stage of injection | date [d/m/y] |
|---|---|---|---|---|---|---|---|---|
| 1 | bituminous | fine | 800 | 1.8 | 13.4 | 60 | lower | 13.05.15 |
| 2 | bituminous | fine | 800 | 4.4 | 35.6 | 50 | lower | 13.05.15 |
| 3 | bituminous | fine | 850 | 1.7 | 12.3 | 40 | lower | 14.04.15 |
| 4 | bituminous | fine | 850 | 2.7 | 20.3 | 20 | lower | 14.04.15 |
| 5 | bituminous | fine | 850 | 2.8 | 20.7 | 30 | lower | 27.05.15 |
| 6 | bituminous | fine | 900 | 4.5 | 33.4 | 15 | lower | 13.05.15 |
| 7 | bituminous | coarse | 800 | 3 | 22.3 | 90 | lower | 08.05.15 |
| 8 | bituminous | coarse | 850 | 3.3 | 24.5 | 105 | lower | 28.07.15 |
| 9 | bituminous | coarse | 900 | 3.3 | 24.5 | 65 | lower | 28.07.15 |
| 10 | lignite | fine | 850 | 4.6 | 28.2 | 40 | lower | 10.10.14 |
| 11 | lignite | fine | 850 | 5.5 | 34.0 | 45 | lower | 10.10.14 |
| 12 | lignite | fine | 900 | 4.6 | 38.6 | 35 | lower | 10.10.14 |
| 13 | lignite | fine | 850 | 3.0 | 18.7 | 30 | lower | 09.09.15 |
| 14 | bituminous | fine | 850 | 2.4 | 17.6 | 100 | lower | 14.01.16 |
| 15 | bituminous | fine | 900 | 2.7 | 20.3 | 70 | lower | 14.01.16 |
| 16 | bituminous | fine | 850 | 2.8 | 20.7 | 30 | **upper** | 23.06.16 |
| 17 | bituminous | coarse | 800 | 3.0 | 22.3 | 20 | **upper** | 14.07.16 |
| 18 | bituminous | coarse | 900 | 3.0 | 22.3 | 25 | **upper** | 14.07.16 |
| 19 | bituminous | coarse | 900 | 5.3 | 38.9 | 10 | **upper** | 14.07.16 |
| 20 | bituminous | fine | 850 | 1.4 | 10.2 | 10 | **upper** | 05.01.17 |
| 21 | bituminous | fine | 850 | 2.4 | 17.8 | 30 | **upper** | 05.01.17 |
| 22 | bituminous | fine | 850 | 3.6 | 26.7 | 20 | **upper** | 05.01.17 |
| 23 | lignite | fine | 850 | 1.7 | 10.3 | 70 | **upper** | 16.01.17 |
| 24 | lignite | fine | 900 | 1.7 | 10.3 | 30 | **upper** | 16.01.17 |
| 25 | lignite | fine | 850 | 1.7 | 10.3 | 30 | **upper** | 16.01.17 |
| 26 | bituminous | coarse | 850 | 2.8 | 21.0 | 40 | **upper** | 15.05.17 |
| 27 | bituminous | coarse | 900 | 2.9 | 21.6 | 20 | **upper** | 15.05.17 |
| 28 | bituminous | coarse | 900 | 3.1 | 23.2 | 15 | **upper** | 15.05.17 |
| 29 | bituminous | coarse | 800 | 3.1 | 23.1 | 50 | **upper** | 29.06.17 |
| 30 | bituminous | coarse | 850 | 3.2 | 23.6 | 30 | **upper** | 29.06.17 |
| 31 | lignite (2) | coarse | 800 | 4.2 | 15.8 | 10 | lower | 20.07.17 |
| 32 | lignite (2) | coarse | 850 | 2.0 | 9.6 | 40 | lower | 20.07.17 |
| 33 | lignite (2) | coarse | 850 | 2.0 | 8.5 | 30 | lower | 20.07.17 |
| 34 | lignite (2) | coarse | 850 | 3.0 | 10.4 | 30 | lower | 20.07.17 |
| 35 | lignite (2) | coarse | 900 | 4.2 | 16.4 | 15 | lower | 20.07.17 |
| 36 | lignite (2) | fine | 840 | 3.9 | 13.8 | 30 | lower | 03.08.17 |
| 37 | methane | - | 850 | 1.1 | 15.3 | 60 | lower | 18.08.17 |
| 38 | lignite (2) | fine | 850 | 3.0 | 10.5 | 160 | lower | 20.09.17 |
| 39 | biomass | coarse | 850 | 2.7 | 14.0 | 90 | lower | 25.01.18 |
| 40 | biomass | fine | 850 | 2.4 | 12.7 | 30 | lower | 03.07.18 |
| 41 | biomass | fine | 850 | 3.1 | 16.0 | 40 | lower | 03.07.18 |

(1): Here only the steady state fuel operation time, which was used for evaluation is listed.
(2): For the later lignite experiments a new batch of raw lignite was used. The water content of the raw lignite was around 50 wt.%.

### 3.1.4. Evaluation of the performance of the CLC pilot plant

The carbon capture rate $\eta_{CC}$ is obtained from the molar flow of carbonaceous gases at the FR outlet minus the molar flow of carbonaceous gases flowing into the FR. This value is divided by the sum of the molar flow of carbonaceous gases exiting from both FR and AR minus the molar flow of carbonaceous gases entering the FR. The carbon capture efficiency $\eta_{CC}$ expresses the carbon capture ability of the CLC system.

$$\eta_{CC} = \frac{[\dot{n}_{CO_2}+\dot{n}_{CO}+\dot{n}_{CH_4}]_{FR_{out}}-[\dot{n}_{CO_2}]_{FR_{in}}}{[\dot{n}_{CO_2}+\dot{n}_{CO}+\dot{n}_{CH_4}]_{FR_{out}}+[\dot{n}_{CO_2}]_{AR_{out}}-[\dot{n}_{CO_2}]_{FR_{in}}} \quad (32)$$

One of the major problems during iG-CLC operation is the conversion of the gasification products, mainly $CH_4$, $H_2$ and CO. This is addressed by the term oxygen demand $\Omega_{OD}$, which describes the oxygen needed to oxidize the not yet converted gases in the exhaust. When gases are not converted and hence their chemical energy is not released, the efficiency of the process is lowered. The oxygen demand $\Omega_{OD}$ is defined by the oxygen needed for complete conversion of the unconverted gases at the FR outlet divided by the molar flow of oxygen needed for complete conversion of all injected fuel. The oxygen demand $\Omega_{OD}$ expresses the amount of oxygen needed to oxidize the combustible component in theexhaust gas into $CO_2$ and $H_2O$ and is calculated based on gaseous components only:

$$\Omega_{OD} = \frac{[0.5\dot{n}_{H_2}+0.5\dot{n}_{CO}+2\dot{n}_{CH_4}]_{FR_{out}}}{\dot{n}_{O_2,min}} \quad (33)$$

with $\dot{n}_{O_2,min}$ being the molar flow of oxygen required for the complete combustion of the fuel. For the research purpose here, the oxygen demand for the lower stage $\Omega_{OD}$ only is separately calculated. This value will show what would have happened in the system if it had only one FR stage. The combustion efficiency of the fuel reactor $\eta_{FR}$ represents the ability of the system to convert fuel inside the FR and it is calculated from:

$$\eta_{FR} = 1 - \frac{[0.5\dot{n}_{H_2}+0.5\dot{n}_{CO}+2\dot{n}_{CH_4}]_{FR_{out}}+\dot{n}_{C,carbon,loss}}{\dot{n}_{O_2,min}-\dot{n}_{C.carbon,slip}} \quad (34)$$

The oxide-oxygen efficiency $\eta_{OO}$is similar to the carbon capture efficiency but is based only on a molar balance of oxygen flowing in, and oxygen and carbon dioxide flowing out of the AR [10]. Since these gases can be easily measured, one can say that this is the more reliable value for the system's carbon capture ability. It is defined as follows:

$$\eta_{OO} = \frac{[\dot{n}_{O_2,in}-\dot{n}_{CO_2,out}-\dot{n}_{O_2,out}]_{AR}}{[\dot{n}_{O_2,in}-\dot{n}_{O_2,out}]_{AR}} \quad (35)$$

The solid fuel conversion $X_C$ denotes the fraction of solid fuel reacting in both FR and AR towards gaseous products and it is defined by:

$$X_c = 1 - \frac{\dot{n}_{C,unconverted}}{\dot{n}_{C,fuel,in}} \quad (36)$$

### 3.1.5. Molar flows inside the pilot plant

The fluidizing velocities ($u_{AR}$, $u_{FR}$, $u_{S1}$, $u_{S2}$) and the gas concentrations ($x_{O_2,AR,out}$, $x_{CO_2,AR,out}$, $x_{CO_2,FR,out}$, $x_{CO,FR,out}$, $x_{CH_4,FR,out}$, $x_{H_2,FR,out}$, $x_{O_2,FR,out}$) from AR and FR outlet are measured. The mass flows of injection gases are arranged with rotameters. Meanwhile, the data is collected and converted to molar flows ($\dot{n}_{O_2,AR,in}$, $\dot{n}_{CO_2,FR,flu,in}$, $\dot{n}_{CO_2,FR,inj,in}$, $\dot{n}_{H_2O,FR,in}$, $\dot{n}_{H_2O,S1,in}$, $\dot{n}_{H_2O,S2,in}$). To investigate the effect of the two stages structure, gas sampling can also be done at the top of the lower stage, before the gases arrive in stage 2. The properties of the gases at the exit of the first stage of FR ($x'_{CO_2,FR,out}$, $x'_{CO,FR,out}$, $x'_{CH_4,FR,out}$, $x'_{H_2,FR,out}$, $x'_{O_2,FR,out}$) can then be used to specify the influence of each single stage. The $N_2$ component in compressed air is regarded to be unchanged in the reactor, and therefore it holds:

$$\dot{n}_{AR,out} \cdot x_{N_2,AR,out} = \dot{n}_{AR,in} \cdot x_{N_2,AR,in} \quad (37)$$
$$\text{with } x_{N_2,AR,out} = 1 - x_{O_2,AR,out} - x_{CO_2,AR,out}$$

With the assumption that the exhaust gas is composed only of the measured gases $CO_2$ and $O_2$, as well as the not measured $N_2$ one can calculate all molar flows exiting the AR. Further, it has to be noted that the local gas concentration around the gas sampling tube is measured, which can deviate from the cross-section average used for the calculations.

Solid carbon is slipping from the FR to the AR. Because of the abundance of oxygen in the AR, the carbon will be completely converted to $CO_2$ and therefore the molar flow of carbon which is lost as a carbon slip can be calculated from:

$$\dot{n}_{C,carbon,slip} = \dot{n}_{AR,out} \cdot x_{CO_2,AR,out} \quad (38)$$

After the operation, the dust in the off-gas of AR and FR is collected in the filter (cf. Figure 14) and the unburnt carbon content is determined from the ignition loss. The ratio of the mass of unburnt carbon collected in the filter to the mass of carbon introduced with the fuel can also be written as ratio of molar flows of carbon and thus defines a carbon loss fraction,

$$y_{C,loss} = \frac{\dot{m}_{C,carbon,loss}}{\dot{m}_{C,fuel,in}} = \frac{\dot{n}_{C,carbon,loss}}{\dot{n}_{C,fuel,in}} = x_{C,loss} \quad (39)$$

To calculate the molar flow of gaseous species at the FR exit, a carbon molar balance is formulated. It holds:

$$\dot{n}_{FR,out} \cdot x_{C\text{ in gas},FR,out} + \dot{n}_{C,carbon,slip} + \dot{n}_{C,carbon,loss} = \dot{n}_{C,fuel,in} + \dot{n}_{C,FR,inj,in} + \dot{n}_{C,FR,flu,in} \quad (40)$$

Where $\dot{n}_{C,FR,flu,in}$ describes the molar flow of C which is introduced for fluidization purpose and $\dot{n}_{C,FR,inj,in}$ denotes the molar flow of C which is used for the injection of the fuel. Rearrangement of equation (40) yields the molar flow at the exit of the FR.

$$\dot{n}_{FR,out} = \frac{\dot{n}_{C,FR,flu,in}+\dot{n}_{C,FR,inj,in}+\dot{n}_{C,fuel,in}-\dot{n}_{C,carbon,slip}-\dot{n}_{C,carbon,loss}}{(x_{CO}+x_{CO_2}+x_{CH_4})_{FR,out}} \tag{41}$$

The molar flow for an individual gas component i can be calculated from the gas concentration measurements for each component $x_i$:

$$\dot{n}_i = \dot{n}_{out} \cdot x_i \tag{42}$$

with index i being CO, $CO_2$, $CH_4$, $O_2$ and $N_2$.

The unknown molar flows from the FR are calculated with a molar balance based on measured outlet concentrations and the input conditions. Furthermore, the dust entrained with the off-gas is examined concerning its content of uncombusted contents. Figure 17 shows the boundaries of the FR calculations.

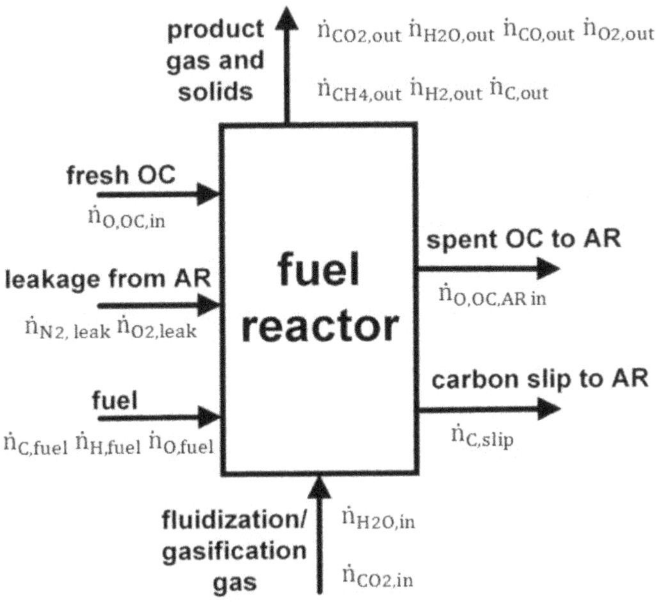


**Figure 17: Molar balancing of the fuel reactor system.**

Both stages from the two-stage system are put together for the molar balancing of the FR. The reason for this procedure is that the solids conversion of the fuel cannot be distinguished between the stages directly. This would have to be evaluated by injection of tracer gases, which was done earlier. A set of linear equations gives molar balances for C, H and O at steady state conditions:

$$(\dot{n}_{C_{fuel}} + \dot{n}_{CO2})_{FR,in} - (\dot{n}_{C_{slip}} + \dot{n}_{CH4} + \dot{n}_{CO2} + \dot{n}_{CO} + \dot{n}_{C_{loss}})_{FR,out} = 0 \tag{43}$$

$$(\dot{n}_{H_{fuel}} + 2\dot{n}_{H2O})_{FR,in} - (4\dot{n}_{CH4} + 2\dot{n}_{H2O} + 2\dot{n}_{H2})_{FR,out} = 0 \quad (44)$$

$$(\dot{n}_{O_{fuel}} + 2\dot{n}_{O2} + \dot{n}_{H2O} + 2\dot{n}_{CO2})_{Fr,in} - (\dot{n}_{CO} + \dot{n}_{H2O} + 2\dot{n}_{CO2} + 2\dot{n}_{O2})_{FR,out} = 0 \quad (45)$$

Measurements in the dry off-gas of $x_{CO2}$, $x_{O2}$, $x_{CO}$, $x_{H2}$, $x_{CH4}$ are used for 5 molar balances for each component i to yield all missing molar streams, shown in exemplary equation (46). This is done according to Thon et al. [15] via a MATLAB routine. From the detailed balances, also the leakage and water streams are calculated. This balance is assuming that all remaining gas, which is not measured, is nitrogen.

$$\dot{n}_{i,out} = x_{i,out}\left(\dot{n}_{N2_{out}} + \dot{n}_{CO_{out}} + \dot{n}_{CH4_{out}} + \dot{n}_{H2_{out}} + \dot{n}_{O2_{out}} + \dot{n}_{CO2_{out}}\right) \quad (46)$$

### 3.1.6. Inventory loss during operation

Flue gases exiting the reactors contain next to their gaseous constituents three major mass flows of particulates, coal ash $m_{ash,dust}$, unconverted carbon $m_{C,dust}$ and abraded OC $m_{OC,dust}$. After the measurement of the gas composition, the flue gas streams from air and FR are mixed and then cleaned from solid material by a filter. All filter cartridges are cleaned automatically in short intervals, so that it can be expected that the fines of one experiment can be assigned to it properly. The ash content is calculated from the amount of coal led into the reactor under the assumption that the ash is completely elutriated from the CLC plant. It is assumed that only the carbon compound of the coal remains in the dust, whereas hydrogen and oxygen leave the reactor in gaseous form. The fixed carbon content is estimated by ignition loss tests at 900°C over a duration of 3 hours. After subtracting the ash $m_{ash,fines}$ and fixed carbon, the remainder resembles the OC, which is abraded and lost during the operational time. The inventory loss was calculated via equation (9) and was based on the total solids inventory $m_{inv}$ of the system.

$$\text{Inventory loss per unit operation time } \left[\frac{w.\%}{h}\right] = \frac{m_{OC,dust}}{t_{operation} \cdot m_{inv}} \quad (47)$$

## 3.2. Experimental lab scale fluidized bed

### 3.2.1. Reactor setup

A laboratory scale bubbling fluidized bed reactor has been operated to investigate the gasification under CLC conditions. In Figure 18, the reactor with 53 mm inner diameter and a height of 500 mm is depicted. It is produced from stainless steel and can be operated at temperatures up to 1050 °C. Fluidization gas is preheated by spirally leading it around the electrically heated furnace. Coal can be fed in-bed and above-bed, by either changing the bed height or the length of the detachable coal feeding pipe.

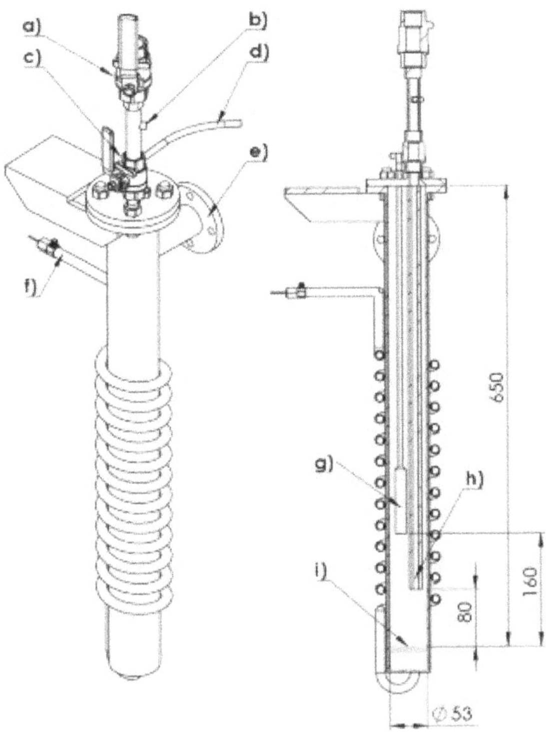


**Figure 18: Laboratory bubbling fluidized bed reactor and its dimensions [mm]. a, c) ball valves to lock samples into the reactor; b) N2 supply to displace ambient air; d) sample gas line; e) exhaust line; f) fluidization gas inlet; g) sample gas filter; h) coal feeding pipe; i) porous plate gas distributor. The reactor is placed into an electrically heated oven, which is not shown here.**

The lab-scale reactor can be operated with $CO_2$, $H_2$, CO, $CH_4$, and $O_2$, air and nitrogen and any blend of them. Water is added into the pipe for the fluidization gas with an electrically controlled syringe. It is then vaporized in the spiral gas preheater. As pictured in Figure 19, the reactor exhaust gas runs through several cleaning steps consisting of tar removal, a condenser, an adsorber filter for nitrous gases and a particle filter, before it is analyzed. $CO_2$, CO, $CH_4$ are analyzed using infrared detection (ir). $O_2$ is detected by the paramagnetic effect (pm). $H_2$ is measured via thermal conductivity (tc) and corrected for the influences of $CH_4$ and $CO_2$ on thermal conductivity.

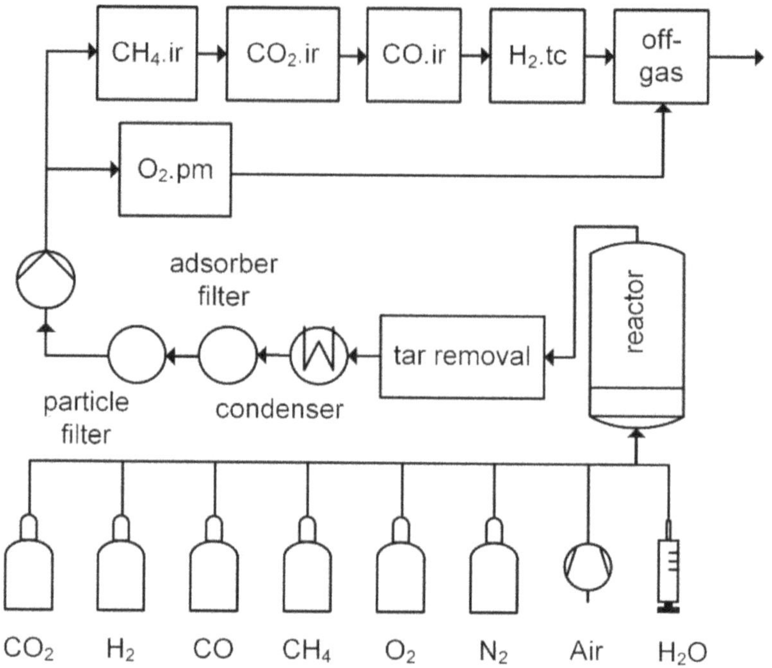


**Figure 19: Process flow sheet of the laboratory reactor system.**

### 3.2.2. Experimental procedure in the laboratory reactor

Char gasification was examined by releasing a certain amount of coal into a fluidized bed reactor, which was fluidized with different mixtures of $N_2$, $CO_2$ and $H_2O$.

Table 10 summarizes all conditions, which were checked inside the fluidized bed reactor. During the gasification experiments, 1g ± 0.005 g of lignite char of different sizes is used and introduced into a fluidized bed of 100 g of quartz sand ($d_p$= 90 - 355 μm, 5 cm fixed bed height). The reactive OC bed experiments were conducted by exchanging the sand bed against 120 - 240 g of OC bed material. The oxygen carrying capacity of the OC was at least 3 - 6 times the amount needed to fully oxidize the gasification products.

An experimental cycle with an inert sand bed starts with the adjustment of the gasification agents' gas flow and the gaseous reactant's concentration. Then the char is introduced into the section between the two ball valves a) and c) of Figure 18. The lower ball valve stays closed, and the upper ball valve is partly open. Then nitrogen is flushed into this section and can exit via the partly opened ball valve. After around one minute the upper ball valve is closed, and the nitrogen flow stopped at the same time. By opening the lower ball valve, the char falls into the bed and the gasification products can be analyzed. When no more gasification products are detected, air is introduced into the reactor, to check, whether all char is converted. When OC is used as bed material, the procedure stays basically the same, because the introduction of air after each experiment will re-oxidize the carrier.

The experiments were assessed as successful, if the carbon balance of the measured exhaust gases was closed within a ± 25% error margin. If the carbon balance could not be closed within this margin, the experiment was repeated, or the results were rejected. 135 out of a total of 169 test runs fulfilled this error margin and were used for further evaluation. After 120 test runs, a series of experiments at 900 °C and 25 vol.-% $CO_2$ was repeated with a new batch of lignite char. The deviation of total time of conversion from the previous runs at the same conditions was below 10%.

All used bed materials (sand and the $Al_2O_3$/CuO-OC) start to fluidize at a velocity of $u_{mf}$ = 0.03 - 0.04 m/s at ambient conditions. The minimal fluidization velocity which will be lower at 800 – 1000 °C. The fluidization gas is adjusted such that without the gasification products a superficial gas velocity $u_R$ under operating conditions of 0.1 - 0.2 m/s is achieved.

**Table 10: Investigated operation conditions inside the laboratory fluidized bed.**

| experimental aim | temperature | $u_R$ | gas concentration (rest $N_2$) | bed | number of char particle size fractions |
|---|---|---|---|---|---|
| $CO_2$ inert bed gasification | 900 °C | 0.15 m/s | 25 vol.-% $CO_2$ | sand | 6 |
| $CO_2$ inert bed gasification | 900 °C | 0.15 m/s | 50 vol.-% $CO_2$ | sand | 6 |
| $H_2O$ inert bed gasification | 900 °C | 0.15 m/s | 25 vol.-% $H_2O$ | sand | 6 |
| $H_2O$ inert bed gasification | 900 °C | 0.15 m/s | 50 vol.-% $H_2O$ | sand | 6 |
| temperature variation $CO_2$ | 1000 °C | 0.165 m/s | 25 vol.-% $CO_2$ | sand | 6 |
| temperature variation $CO_2$ | 1000 °C | 0.165 m/s | 50 vol.-% $CO_2$ | sand | 6 |
| temperature variation $H_2O$ | 1000 °C | 0.165 m/s | 25 vol.-% $H_2O$ | sand | 6 |
| temperature variation $H_2O$ | 1000 °C | 0.165 m/s | 50 vol.-% $H_2O$ | sand | 6 |
| temperature variation $CO_2$ | 800 °C | 0.135 m/s | 25 vol.-% $CO_2$ | sand | 6 |
| temperature variation $CO_2$ | 800 °C | 0.135 m/s | 50 vol.-% $CO_2$ | sand | 6 |
| temperature variation $H_2O$ | 800 °C | 0.135 m/s | 25 vol.-% $H_2O$ | sand | 6 |
| temperature variation $H_2O$ | 800 °C | 0.135 m/s | 50 vol.-% $H_2O$ | sand | 6 |
| $CO_2$ reactive bed gasification | 900 °C | 0.15 m/s | 25 vol.-% $CO_2$ | OC | 6 |
| $CO_2$ gasification inhibition CO | 900 °C | 0.15 m/s | 25 vol.-% $CO_2$ + 10 vol.-% CO | sand | 6 |
| $CO_2$ gasification inhibition $H_2$ | 900 °C | 0.15 m/s | 25 vol.-% $CO_2$ + 15 vol.-% $H_2$ | sand | 6 |
| simultaneous $H_2O$ and $CO_2$ gasification, inert bed | 900 °C | 0.15 m/s | 25 vol.-% $CO_2$ + 25 vol.-% $H_2O$ | sand | 3 |
| temperature variation, simultaneous $H_2O$ and $CO_2$ gasification | 800 °C | 0.135 m/s | 25 vol.-% $CO_2$ + 25 vol.-% $H_2O$ | sand | 3 |
| temperature variation, simultaneous $H_2O$ and $CO_2$ gasification | 1000 °C | 0.165 m/s | 25 vol.-% $CO_2$ + 25 vol.-% $H_2O$ | sand | 3 |
| variation of $H_2O$ and $CO_2$ concentration | 900 °C | 0.15 m/s | 15 vol.-% $CO_2$ + 35 vol.-% $H_2O$ | sand | 3 |
| variation of $H_2O$ and $CO_2$ concentration | 900 °C | 0.15 m/s | 15 vol.-% $CO_2$ + 35 vol.-% $H_2O$ | sand | 3 |

**Table 10 (continued)**

| | | | | | |
|---|---|---|---|---|---|
| $H_2O$ reactive bed gasification | 900°C | 0.15 m/s | 25 vol.-% $H_2O$ | OC | 3 |
| simultaneous $H_2O$ and CO2 gasification, reactive bed | 800 °C | 0.135 m/s | 25 vol.-% $CO_2$ + 25 vol.-% $H_2O$ | OC | 3 |
| temperature variation, simultaneous $H_2O$ and $CO_2$ gasification | 1000 °C | 0.165 m/s | 25 vol.-% $CO_2$ + 25 vol.-% $H_2O$ | OC | 3 |
| temperature variation, simultaneous $H_2O$ and $CO_2$ gasification | 900 °C | 0.15 m/s | 15 vol.-% $CO_2$ + 35 vol.-% $H_2O$ | OC | 3 |
| variation of $u_R$ | 900 °C | 0.10 m/s | 25 vol.-% $CO_2$ + 25 vol.-% $H_2O$ | OC | 3 |
| variation of $u_R$ | 900 °C | 0.20 m/s | 25 vol.-% $CO_2$ + 25 vol.-% $H_2O$ | OC | 3 |
| variation of $H_2O$ and $CO_2$ concentration | 900 °C | 0.15 m/s | 15 vol.-% $CO_2$ + 35 vol.-% $H_2O$ | OC | 3 |
| variation of $H_2O$ and $CO_2$ concentration | 900 °C | 0.15 m/s | 15 vol.-% $CO_2$ + 35 vol.-% $H_2O$ | OC | 3 |
| Reduced reactive bed height 240 g - > 120 g | 900 °C | 0.15 m/s | 25 vol.-% $CO_2$ | OC | 3 |
| Reduced reactive bed height 240 g - > 120 g | 900 °C | 0.15 m/s | 25 vol.-% $H_2O$ | OC | 3 |

### 3.2.3. Data evaluation in the laboratory reactor

Molar balance from gas concentration measurements

The rate of gasification, equation (48), relates the weight change of char over time $\frac{dw_C}{dt}$ to the amount of unreacted char at the beginning of the measurement $w_{C,0}$:

$$R_g(X_S) = -\frac{1}{w_{C,0}}\frac{dw_C}{dt} = \frac{dX_S}{dt} \tag{48}$$

The weight change of the char inside the fluidized bed reactor is difficult to be measured directly, like it is done for example in a thermogravimetric analyzer (TGA). Therefore, the analysis is based here on the composition of the gases exiting from the reactor system. The amount of char that has been gasified in a certain time span can be integrated from the total volumetric gas flows $F_{in}$ and $F_{out}$ and the measured volumetric gas concentrations of carbonaceous gases of inlets $x_{C,in}$ and outlets $x_{C,out}$ (t):

$$w_C(t) = \frac{p \cdot M_C}{R \cdot T} \int_0^t (F_{out}(t) \cdot x_{C,out}(t) - F_{in} \cdot x_{C,in})\, dt \tag{49}$$

The measured carbonaceous gases consist of methane $x_{CH4}$, carbon dioxide $x_{CO2}$ and carbon monoxide $x_{CO}$. The inlet gas volume flow $F_{in}$ is adjusted by rotameters. The outlet gas volumetric flow rate $F_{out}$ (t) is calculated from equations (50) and (51) with the assumption that reactions of nitrogen can be neglected:

$$F_{out} = \frac{F_{N2,in}}{x_{N2,out}} \tag{50}$$

$$x_{N2,out} = 1 - x_{CH4,out} - x_{CO2,out} - x_{CO,out} - x_{H2,out} - x_{O2,out} \tag{51}$$

Fitting the experimental data to particle reaction models

With the molar balancing, it is possible to calculate the rate of gasification for every measured time step of the reactor setup. The experimental progress of the char conversion can also be made visible with this data. In a next step, the experimental data is compared to different particle reaction models. As described in Section 2.3, the Homogeneous or Volumetric Reaction Model (VRM) and the Shrinking Core Model (SCM) are investigated, whether the observed reaction pattern can be described accurately with those. The rate of gasification $R_g(X_s)$ is experimentally determined for a given temperature T and a given reactive gas species g and then modeled with the particle reaction models $f_{model}(X_s)$:

$$R_g(X_s) = \frac{dX_S}{dt} = k_{model}(T) \cdot C_g{}^n \cdot f_{model}(X_S) = \frac{1}{\tau(T, C_g{}^n)} \cdot f_{model}(X_S) \tag{52}$$

Here, $k_{model}$ is a kinetic parameter for a certain set of operation conditions, e.g. temperature T and reactive gas molar fraction $x_i$. The kinetic parameter can also be expressed as the reciprocal value of the total time of conversion $\tau$, which can be experimentally determined. The term $f_{model}(X_s)$ is different for the different particle reaction models, namely $f_{VRM}$ and $f_{SCM}$, respectively. If the investigated particles react according to the VRM, a linear dependency of the rate of reaction with the remaining amount of reactive material, here the lignite char, is observed. The VRM supposes a reaction, which takes place over the whole volume of the particle. Following Kunii and Levenspiel [25] it holds:

$$f_{VRM}(X_s) = (1 - X_S) \tag{53}$$

The SCM, as explained by Kunii and Levenspiel [25], describes the solid reaction as taking place on the outer surface of an unreacted core. The core is shrinking during the solid reaction and, hence, the reactive surface as well, while an ash layer of increasing depth is formed by conversion of the char. The reaction rate is limited by resistances, namely the film diffusion through the film layer around the particle, the ash diffusion through the ash deposit around the unreacted shrinking core and a general reaction limitation at the reactive surface. For a reaction that is controlled by the chemical reaction at the active surface the function term then becomes:

$$f_{SCM}(X_s) = (1 - X_S)^{\frac{2}{3}} \tag{54}$$

All the conversion data from the gas concentration measurements and, with it, the rate of gasification was modeled with the SCM and VRM model, respectively. Problems arise when it has to be judged, which reaction model describes the conversion behavior the best. Simply plotting the conversion over time for the experiments and the different models will yield little differences among the models and the experiments. Therefore, some research groups use so called linearized model equations. Fermoso et al. [115] were evaluating steam gasification for 5 solid fuel chars, whereas Tomaszewicz et al. [116] made a kinetic analysis of lignite and two sub-bituminous coals in $CO_2$ atmosphere. The model linearization yields directly the kinetic constant $k_{model}$ for certain operation conditions, which is the slope of the graphs shown in Figure 20. It also shows the deviation from a perfect particle model behavior. In Figure 20, the experimental data is plotted in the linearized form of the three particle reaction models.

The linearized form is taken from integration of the rate expression equation (52) with the different function terms of the models in equations (53) - (54) and is shown in the following equations.

VRM: $$-\ln(1-X_s) = k_{VRM}\, t \tag{55}$$

SCM: $$3\left[1-(1-X_S)^{\frac{1}{3}}\right] = k_{SCM} t \tag{56}$$

In the exemplary plots in Figure 20 it becomes visible that the gasification can best be described with a SCM behavior, because the slope deviates the least from the linear one.

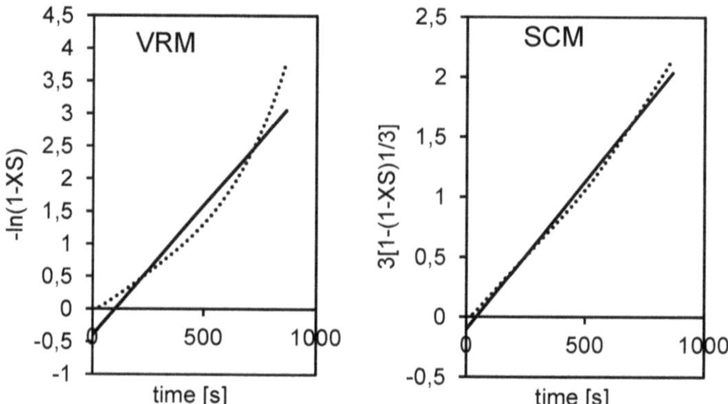


**Figure 20: Linearized plots of the experimental data at 900 °C, 25 vol.-% $_{CO2}$, $d_S$ = 1215 µm. Experimental data (broken lines) compared to ideal model behavior (solid lines).**

Another approach to estimate the accuracy of the used model is to divide the gasification rate by the rate expression $f_{model}$ ($X_s$) itself, shown in equation (52), used by Cuadrat et al. [93], Brown et al. [96] and described by Johnson et al. [88]. This approach will turn out the instantaneous gasification rate $R_{G,inst}$ ($X_s$). This instantaneous rate connects the gasification rate $R_g(X_s)$ with the remaining amount of char present via the particle reaction model according to:

$$R_{G,inst}(X_s) = \frac{R_G(X_S)}{f_{model}(X_S)} = k_{model} \tag{57}$$

If the instantaneous rate of gasification can be plotted as a constant value over most of the conversion, it can be stated that the gasification follows the chosen particle reaction model $f_{model}$ ($X_s$). This can be seen in Figure 21, where the instantaneous rate is plotted. The instantaneous rate and thus $k_{model}$ is relatively constant at a value around 0.003 1/s for the SCM. The VRM model shows a quite strong change of $k_{model}$ with $X_S$, which makes this model less suitable for the description of this case.

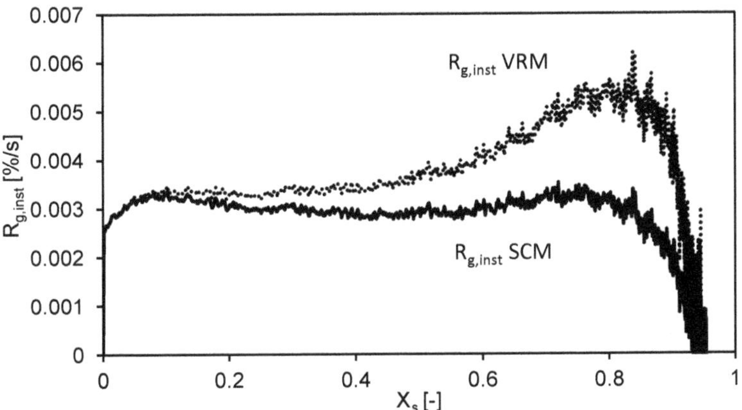


**Figure 21: Exemplary instantaneous rate of gasification for a gasification experiment at 900 °C, 25 vol.-% $CO_2$, $d_S$ = 752 µm.**

Kinetic parameters extraction

The ultimate goal of the reaction kinetic analysis is to find a general expression for the kinetic parameter $k_{model}$. In the case of chemical reaction control, the parameter can then be described with an Arrhenius type kinetics:

$$k_{model}(T, C_g) = k_0 \cdot e^{-\frac{E_a}{RT}} \cdot C_g^n \tag{58}$$

In equation (58), $k_0$ is the pre-exponential factor, $E_a$ the activation energy and R the universal gas constant. Further, $C_g$ is the molar concentration of the reacting gas and n the reaction order. As in this work a broad spectrum of particle sizes has been investigated, it is probable that for larger particle sizes the reaction kinetics is also dependent on diffusive mass transfer phenomena, which will limit the progress of conversion. Levenspiel [68] described thoroughly how the different phenomena can have an effect on the conversion of particles with shrinking-core behavior and proposed for the combustion of pure carbon:

$$\frac{1}{k_{model}(T,C_g^n,d_p)} = \frac{1}{k_{rc}(T,C_g^n)} + \frac{1}{k_g(d_S)} \tag{59}$$

In this latter equation $\frac{1}{k_{rc}(T,C_g^n)}$ denotes the resistance due to the surface reaction and $\frac{1}{k_g(d_S)}$ the resistance due to diffusion through the gas film layer and the developing ash layer around the core of the particle. This concept divides the modeled reaction rate into a part that is dependent on the particle size and one part that is dependent on the temperature and gas concentration. The diffusion resistance can be approximated with an effective diffusivity $D_e$ ($m^2/s$):

$$k_g(d_p) = \frac{D_e}{\left(\frac{d_p}{2}\right)^2} \tag{60}$$

The presently used bubbling fluidized bed reactor makes it possible to investigate the gasification of char in the presence of reactive OC particles. We make the simplifying assumptions that

- the gas concentrations along the reactor height are constant
- the oxidation state of the reactive OC is not changing during a gasification test run
- the gas bypass in form of bubbles is neglected

With these assumptions kinetic parameters can be extracted as a first approximation.

## 3.3. Dynamic Flowsheet Simulation

The programming work by Skorych et al. [117] and the modeling work presented here were conducted within the Priority Program of the German Research Foundation (DFG), called SPP 1679, "Dynamic Simulation of Interconnected Solids Processes". The present work was done in course of the sub-project: "Dynamic Simulation of Interconnected Fluidized Bed Reactors". The core of this project is the development of dynamic process models for a flowsheet simulation environment called Dyssol, which integrates all parts that are developed during the priority program. The Dyssol software follows a sequential-modular approach, which means that each process unit is modeled and calculated separately and in a sequential order. Due to the presence of backflows and recycle streams, which can affect preceding units, the whole flowsheet must be iterated, until the error of the calculated data in the tear streams is within a certain error margin.

The aim of the present Ph.D. project is to describe the hydrodynamic situation of all components inside a CLC system of interconnected fluidized bed reactors. With it, from only few input parameters, i.e. solids inventory and fluidization velocities, the behavior of the system should be predicted. In addition to that, this work considers the dynamic changes of solid circulation and bed mass as a result from changing the operation conditions. After the simulation of the hydrodynamics of the system, a chemical reaction mechanism based on a two-phase model is used to calculate the chemical conversion of fuel and OC.

To do that, all components, which comprise a chemical looping facility and their effects on each other, were implemented into the novel flowsheet simulation environment Dyssol whose core elements can be found in Dosta et al. [118]. The main features were extended by Skorych et al. [117]. This simulation environment offers the functionality to capture dynamic effects within the whole system of a CLC plant. The required models span from the fluidized bed reactors, in bubbling and circulating operation to loop seals and the cyclone.

The Dyssol software was used to generate a flowsheet for a CLC system. In a first approach the focus is put on hydrodynamic models for the AR (a circulating fluidized bed), the FR stages (two bubbling beds), the cyclone and the loop seals, which are all included into Dyssol.

Process flowsheets comprise the main process units, which are linked to share their relevant information with each other via process streams. CLC is carried out here in a system of interconnected fluidized bed reactors. These reactors are separated by loop seals to prevent gas leaking. The solids are separated from the gas exiting the AR with a cyclone. When the fluidized bed reactor is operated in a bubbling regime a cyclone is not essential as a first approach. Figure 22 shows how the experimental facility is transferred into the flowsheet. The network of the main process units from a CLC system is shown: the AR, the two FR stages

(FR1 + FR2), the cyclone (C) and the loop seals, working as siphons (S1 + S2). The streams shown in the Figure 22 bear information about the mass flows of gas, solid and liquid phases, chemical composition, particle size distribution and temperature. Stream (1) starts from the outlet of the AR and enters the cyclone.

In the cyclone the stream is divided by a model into an underflow of coarse particles (3) and an overflow with air and fine solids (2), which leave the reactor system. Particles in S1 are fluidized by fluidization gas stream (5). The siphon will have an overflow (4) which is dependent on the height of solids above the gas distributor inside the unit. In the second FR stage the particles are fluidized by the off-gas of the first stage (7). The outlet of the FR (6) consists of the gases formed in the stage and the fluidization gas, as well as entrained particles. When the overflow height is reached inside FR stage 2, an overflow (8) is formed, which represents the solids flow through the standpipe. FR 1, itself, is fluidized by a mixture of fluidization gas and fuel, which is methane in the simulated case. Coal combustion would be modeled via a separated stream entering the FR stage 1 at a certain height above the distributor. Parts of the fluidization gas of S2 rise via the standpipe and contribute to fluidization of FR1. Solids and gases from S2 recirculate back to the AR. The whole flowsheet is calculated iteratively to solve the whole loop including multiple backflows in the unit.

For the flowsheet simulation, unit modules, which contain the process models and further correlations for the chemical conversion, had to be developed and introduced into the flowsheet software. The used models, fluidized bed reactor, cyclone and loop seal are described in the next section. With these unit modules inside the process flowsheet, the whole process and the chemical reactions in the reactors can be calculated. The OC's reduction by fuel gas $CH_4$ and its decomposition products CO and $H_2$ takes places in FR 1 and FR 2. Further, in the AR, the oxidation of OC with $O_2$ from the fluidization air is included in the simulation.

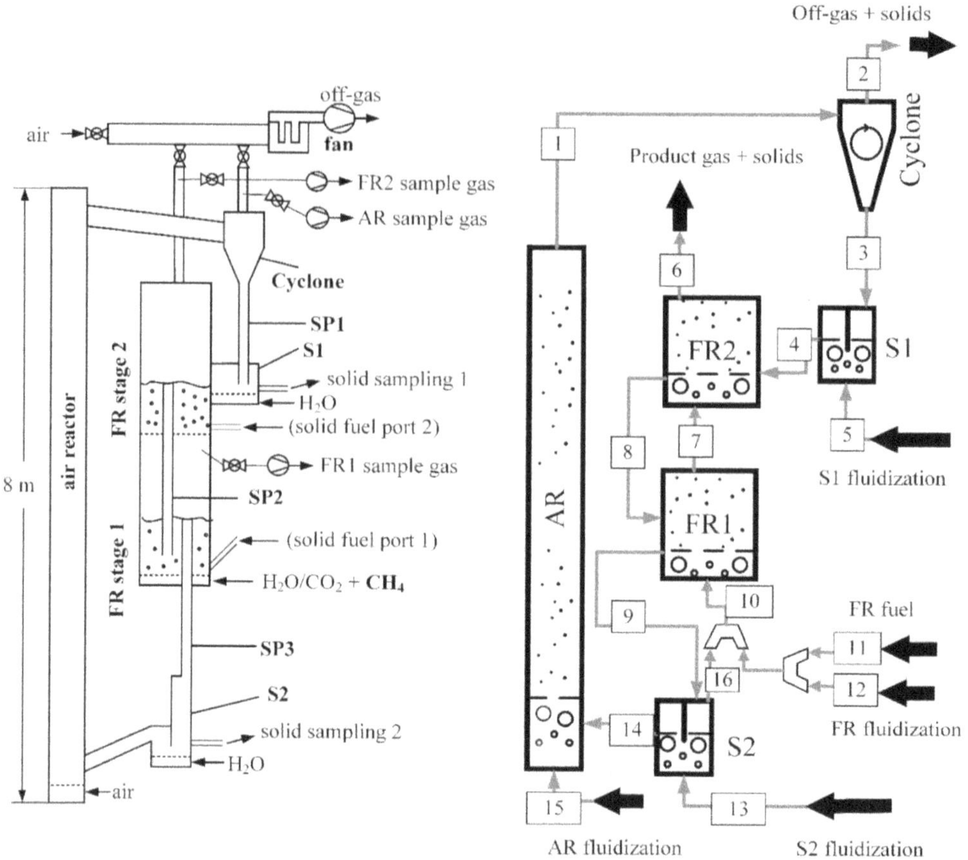


**Figure 22: CLC plant schematics and process flowsheet for simulation.**

All process units except for the cyclone were modelled with a time dependence in their hydrodynamics. The time dependence is expressed by the mass holdup inside each unit and mass fluxes from and towards them, which can change over time. For the cyclone modeling, a steady state approach was used, because the mass hold-up inside it is very small compared to the fluidized bed reactors and siphons.

When two units in a flowsheet relate to each other via streams in both directions the streams are called tear streams. Such streams are harder to calculate than streams from units, which are connected sequentially, because an iterative solution must be found. In the CLC process generally, the back flux of solids to the AR (14) is a tear stream. In the observed system, FR2 overflow (8) is also a tear stream, as both stages' hydrodynamics are directly dependent on each other.

### 3.3.1. Hydrodynamic modeling of fluidized beds

For all three fluidized bed reactors a single module was developed, which considers the solid free bubbles and a solid dense suspension phase [30]. In the module, the reactor is described by two zones: a solid rich dense bottom zone and an upper region with a lower solids' concentration, which is exponentially shrinking towards the top height of the reactor. Inside the bottom zone, the presence of bubbles is assumed for all fluidization velocities. Own research has shown that, also for circulating fluidized bed reactors, the presence of a dense bottom zone can be detected [119]. The correlations used for the description of the fluid mechanics are shortly summarized below. More details can be found in earlier publications [120].The initial bubble size $d_{v,0}$, expressed as the bubble volume equivalent sphere diameter in the dense bed, is given by Davidson and Harrison in the book of Gilliland [121]:

$$d_{v,0} = 1.3 * \left(\frac{V_{or}^{2}}{g}\right)^{0.2} \tag{61}$$

where $V_{or}$ denotes the volumetric flow through a single orifice. With increasing height above the distributor, the bubble volume equivalent diameter $d_v$ is described by Werther and Wein [120] as:

(62)

with $\varepsilon_b$ being the bubble volume fraction and $\lambda$ the average bubble lifetime. The formed bubbles rise with the velocity $u_b$:

$$u_b = \dot{V}_b + 0.71 \cdot \theta \cdot \sqrt{g * d_v} \tag{63}$$

In this equation, $\dot{V}_b$ is the visible bubble volumetric flow and $\theta$ denotes a scale dependent geometry parameter. Within the dense bottom zone, solids can only be found in the suspension phase, as the bubbles are assumed to be solid free. The concentration of solids in the suspension phase $c_{v,suspension}$ is assumed to be close to the concentration at minimum fluidization $c_{v,mf}$. At the border between dense bottom and dilute upper zone, at height $H_b$, the solids volume concentration $c_v(H_b)$ average over the bubble and suspension phases is calculated:

$$c_v(H_b) = (1 - \varepsilon_b(H_b)) \cdot c_{v,suspension} \qquad \text{for } h < H_b \tag{64}$$

Above the dense bottom zone, a correlation, which assumes an exponential decay of solids concentration towards the reactor top is used (Kunii and Levenspiel, 1991). Furthermore, an elutriation effect that differs for each particle size class i is taken into consideration:

$$c_{v,i}(h) = c_{v,i,\infty} + \left(c_{v,i}(H_b) - c_{v,i,\infty}\right) \cdot e^{-a(h-H_b)} \qquad \text{for } h \geq H_b \tag{65}$$

Kunii and Levenspiel [25] describe that the decay factor multiplied by the superficial gas velocity u is taking a constant value for a certain system. The exponential decay factor a is dependent on the superficial gas velocity u and using information from Kunii and Levenspiel it holds,

$$a \cdot u_R = 0.5 \ldots 3 \qquad \text{for bubbling beds and } d_p \sim 300\ \mu\text{m} \tag{66}$$

$$a \cdot u_R = 2 \ldots 12 \qquad \text{for turbulent fluidized beds} \tag{67}$$

The index i in equation (65) denotes that the calculation is done for every single particle class. This leads to an elutriation effect due to different particle sizes and hence varying fluid mechanic properties. The simulation environment Dyssol can handle particle size distributions (PSD) and any transformation and size changes within.

The concentration of solids $c_{v,i,\infty}$ at transport disengagement height (TDH) is calculated from,

$$c_{v,i,\infty} = \frac{G_{s,i,\infty}}{\rho_s \cdot u_R} \tag{68}$$

The solids circulation rate $G_{s,i,\infty}$ for every size fraction i is calculated from an elutriation rate $K_{i,\infty}$ and the mass of particles in each respective class $\Delta Q_{3,i}$,

$$G_{s,i,\infty} = \Delta Q_{3,i} \cdot K_{i,\infty} \tag{69}$$

Correlations for $K_{i,\infty}$ were examined for a wide range of operating conditions, described for example by Chew et al. [122]. Tasirin and Geldart [123] proposed a correlation for bubbling bed operation equation (70). For higher gas flows the elutriation rate constant $K_{i,\infty}$ according to Choi et al. [124] from equation (71) is applied. The correlations were chosen, because they were experimentally elaborated from conditions close to the ones expected in the reactor system of this work. Due to their empirical origin, the equations lack correct units and dimensions. In these equations $\rho_g$ is used as the gas density in kg/m³, $u_{t,i}$ the particle's terminal velocity in m/s for each size interval i, $Ar_i$ being the Archimedes number for each size interval i, $\mu$ is the dynamic viscosity of the gas in Pa s, $F_{g,i}$ and $F_{d,i}$ denote the gravitational and drag force respectively, which are also influenced by particle size. The impact of gravitation and particle drag were experimentally fitted, so the dimensions of the units must adapted to match the final units of Ki, ∞.

$$K_{i,\infty} = 14.5 \cdot \rho_g \cdot u_R^{2,5} \cdot \exp(-5.4 \cdot \frac{u_{t,i}}{u_R}) \tag{70}$$

$$\frac{K_{i,\infty} \cdot d_{i,p}}{\mu} = {Ar_i}^{0.5} \cdot \exp(6.92 - 2.11 \cdot {F_{g,i}}^{0.303} - \frac{13.1}{{F_{d,i}}^{0.902}}) \tag{71}$$

The environment Dyssol allows for a quick change of these correlations. Implementation was carried out in a way that reactors with different hydrodynamic situations could be described with the same module.

The solids mass $m_r$ inside the reactor is obtained for every time step via a mass balance over the upper dilute and the dense bottom zone. $H_r$ is the total height of the reactor and $A_r$ denotes its cross-sectional area. $H_b$ gives the height of the dense zone.

$$m_r = A_r \cdot (\int_0^{H_b} c_v(h)\, dh + \int_{H_b}^{H_r} c_v(h)\, dh) \cdot \rho_s \tag{72}$$

The module distinguishes between bubbling and fast fluidized bed operating condition in the way particles exit the reactor. In a bubbling fluidized bed reactor, solid overflow over the weir is the predominant solid discharge and only little elutriation takes place. It was modeled with a correlation adapted from Botsio and Basu [125] who describe the flow over an overflow weir with an adapted equation of the sharp-crested weir theory by Roberson and Crowe [126]:

$$\dot{m}_{overflow} = \frac{2}{3} C_{weir} \cdot w_{weir} \sqrt{2g} \cdot h_{above}^{\frac{2}{3}} \tag{73}$$

In the equation C is a weir coefficient, W the width of the weir/standpipe and $h_{above}$, the height of the formed bed formed higher than the height of the standpipe or weir $H_{SP}$. When simulating a circulating fluidized bed riser reactor, the mass flow $\dot{m}_{overflow}$ is switched off and all the particles exiting the reactor are stemming from elutriation. The number of elutriated particles at the top of the reactor is dependent on the solids concentration there and the gas velocity, whereas the mass flow over the weir depends on the height of the dense bubbling bed above the weir. For the both operation states, bubbling and fast fluidized, this means that the mass outflow of particles by elutriation $\dot{m}_{E,i}$ is determined by the concentration $c_{v,i}$ of elutriated particles for each size class i at the top of the reactor $H_r$, the reactor cross section $A_r$, the velocity u there and the solids density $\rho_s$,

$$\dot{m}_{elutriation} = \sum c_{v,i}(H_r) \cdot \rho_s \cdot A_r \cdot u_R \tag{74}$$

In steady state calculations fluidized bed reactors would have a certain constant bed mass and inflow of solids would be equal to the outflow. For dynamic simulations fluidized bed reactors have to be described in a transient way, so that accumulation or loss of bed mass is possible. The CFB riser reactor has an inherent outflow defined by the concentration of solids at the maximum height of the reactor. Together with a certain inflow provided by other equipment inside the system, the CFB riser reactor and the bed mass within is sufficiently modelled dynamically. In the bubbling bed reactor the bed height will increase or decrease until the height of the overflow weir or standpipe, denoted as $H_{SP}$, is reached. This means that below a certain bed height, only few particles will leave the reactor due to entrainment. Until the overflow height is reached, no overflow will occur, and an accumulation will take place. Contrarily, if the actual bed height is above $H_{SP}$, an increased overflow would take place so that the bed height reaches the overflow or standpipe height after some time.

In the present status of the simulation environment, the changes in the hydrodynamics, for example because of changes in the fluidization gas velocity, are assumed to occur instantaneously. It is only the change of the solid masses inside the fluidized bed reactors, the loop seals and standpipes which are modelled dynamically. This means that for every time point a quasi-steady state solution is calculated, which determines the outflows of the reactor. The change of the mass inside the bubbling bed reactor is then defined by the difference of mass flow of solids entering the reactor $\dot{m}_{solids,in}$ and the mass flows leaving the reactor by entrainment $\dot{m}_E$ and the overflow of the standpipe $\dot{m}_{SP}$:

$$\frac{dm_r}{dt} = \dot{m}_{solids,in} - \dot{m}_{elutriation} - \dot{m}_{overflow} \tag{75}$$

For the calculation, the reactor is discretized in a certain number of height units. For each of the height units the set of equation for the bubble size or the exponential decay of solids concentration is solved. The reactor is discretized into height elements, typically 500 for the FRs and 1500 for the AR, which was sufficient for the solver to reach a solution within a certain error margin. The convergence criterion within the fluidized bed reactor model is that the total bed mass is expressed by the dilute and dense phase bed mass within the absolute error margin of $10^{-3}$ kg and relative error of $10^{-5}$.

### 3.3.2. Reaction kinetics in fluidized bed

In the fluid mechanics section above, it was mentioned that in the bottom zone, two phases are considered, a solid free bubble phase and a gas-solid suspension phase. Figure 23 illustrates how this affects the reactions and how they were modeled in general.

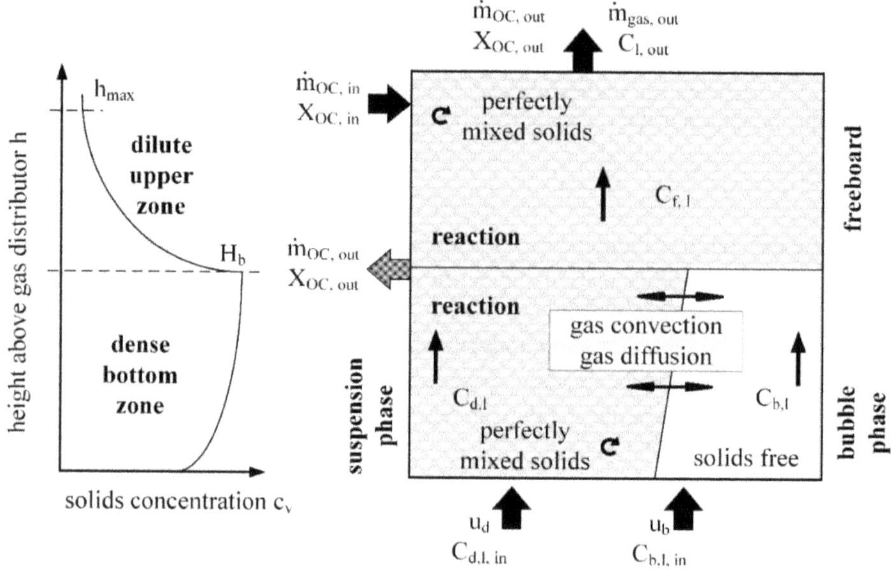


**Figure 23: Left: typical solids distribution inside a fluidized bed reactor. Right: input and output flows inside a fluidized bed reactor, as well as the flows and distributions inside.**

Heterogeneous gas-solid reactions can only to take place inside the suspension phase, and only gas phase reactions can occur inside the bubbles. In the dilute upper region of the reactor, the solids and gases are assumed to form a dilute suspension. The gases are assumed to rise up from the distributor, and back mixing does not occur. Hence, the gas conversion of each species is modeled as in a plug flow reactor. The solids are considered to be perfectly mixed inside the whole fluidized bed reactor.

The oxygen carrier used at TUHH is very similar to the one investigated by the researchers at CSIC Zaragoza using Puralox NWa-155 as the base material and impregnation as the production method. Therefore, in this work, at first the reaction kinetics described in literature, which were determined from TGA experiments are used to describe the gas conversion [69,78]. In a second simulation, the reaction kinetics were varied in a way to fit to the measured gas concentrations at the outlet of both FR stages. The reaction module is designed to allow for different solid reaction models as well as other reaction rates.

The reactions of equations (22) - (25) were considered for the reduction and oxidation of the oxygen carrier in the FR. The reaction of the solid oxygen carrier j with its respective gas component l is described by the reaction rate $r_{s,j,l}$:

$$r_{s,j,l} = -c_{v,j} \cdot \rho_{m,j} \cdot \frac{dX_{s,j,l}}{dt} \tag{76}$$

In this equation, $c_{v,j}$ denotes the volumetric concentration of solid reactant j in the reactor volume element. It is taken by multiplying the total solids concentrations $c_v$ and the fraction of active reactant, CuO, on and inside the particles. Further, $\rho_{m,j}$ is the molar density of the solids reactant j and $\frac{dX_{s,j,l}}{dt}$ is the -solids conversion rate of reactant j with respect to fuel gas l. The fuel gases included in the simulations are $CH_4$, CO and $H_2$. The reaction rate for these gases $r_{g,j,l}$ can be determined by multiplying the solid conversion rate $r_{s,j,l}$ with the stochiometric factor stemming from equations (22) - (25).

The solids conversion rate $\frac{dX_{s,j,l}}{dt}$ is usually determined by experimentally exposing the oxygen carrier with the active compound j to different reacting gases l at various concentrations and temperatures. From this data, Arrhenius type reaction rates can be derived. Several heterogeneous reaction models were proposed to capture the solids conversion, because OCs of different metallic compounds differ in their structure and, hence, also in their reaction behavior. Abad et al. [78] concluded that a shrinking-core-model, which takes the existence of reactive grains into consideration, is well suited to describe the performance of a $CuO/Al_2O_3$ OC. The general form of reaction kinetics that the Arrhenius models are based on is shown:

$$\frac{dX_{s,j,l}}{dt} = k(C_l{}^n, T) \cdot f(X_s) \tag{77}$$

There k is the kinetic constant, which describes the rate of solid conversion as a function of the molar concentration of the reacting gas $C_l$, the order of reaction n and the temperature T and is multiplied by the model function of the solids conversion $X_S$. Assuming that the concentration $C_l$ is constant throughout the course of the reaction in a discretized volume element $A_r \cdot dh$, the kinetic constant k can be described by an Arrhenius type equation (14). In the Arrhenius equation, $k_0$ is a pre-exponential factor and $E_a$ is the activation energy for the reaction, both are determined experimentally. The rate of conversion is also a function of the conversion $X_s$ itself. This is often expressed by an algebraic approach. In the present work an algebraic expression for spherical grains and a reaction limitation is used as described by Adanez et al. [10] for the reaction of the OC:

(78)

The molar balances in the discretized volume elements $A_r \cdot dh$ for the gas species are shown in Figure 24. One can see that in the present case of gas-solid reactions no homogeneous reaction takes place in the bubble phase. Changes in the molar concentrations inside the bubble phase are solely due to the diffusive and convective mass transfer. The heterogeneous reactions in the dense suspension phase will induce a concentration gradient and, with it, diffusive and convection mass transfer.

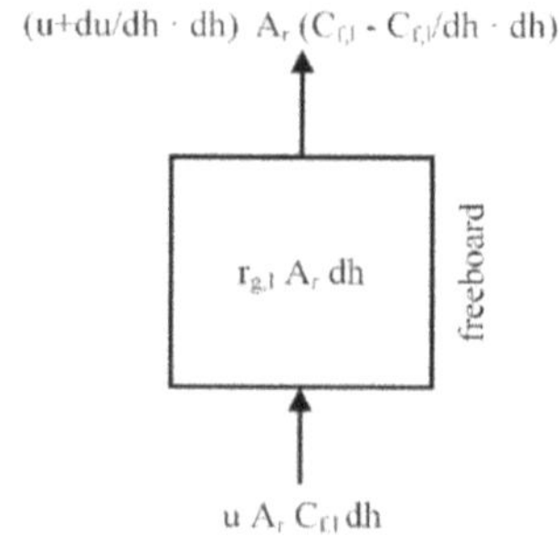


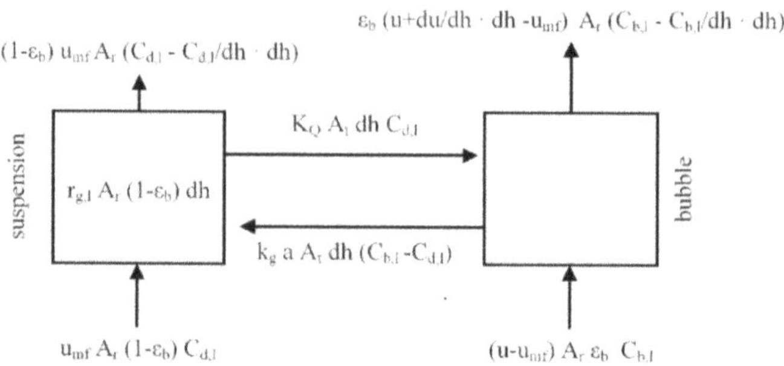


**Figure 24: Molar balances to be solved for each discretized height element dh. The suspension and the bubble phase are connected via diffusive and convective mass transfer.**

The concentration changes of all gas species l at a reactor volume element $A_r \cdot dh$ are given for the bubble phase b, the suspension phase d and the freeboard f:

$$\frac{dC_{b,l}}{dh} = \frac{\dot{j}_{Q,l} - k_G \cdot a \cdot (C_{b,l} - C_{d,l}) - C_{b,l}\frac{du}{dh}}{u - u_{mf}(1-\varepsilon_b)} \tag{79}$$

$$\frac{dC_{d,l}}{dh} = \frac{\dot{j}_{Q,l} + k_G \cdot a \cdot (C_{b,l} - C_{d,l}) + r_{g,l}}{u_{mf}(1-\varepsilon_b)} \tag{80}$$

$$\frac{dC_{f,l}}{dh} = \frac{r_{g,l} - C_{f,l}\frac{du}{dh}}{u_R} \tag{81}$$

In these equations, $r_{g,l}$ denotes the reaction rate of the gas l, induced by the solid oxygen carrier and based on the reactor volume. Homogeneous gas-gas reactions could also be included here, but the chemical reaction scheme described above does not include these reactions. Therefore, only heterogeneous gas-solid reactions will influence the gas concentrations. The chemical reactions inside the reactors have an effect on the total molar flow of gases and, with it, the gas velocity will be changed. Since the velocity inside the suspension phase is assumed to be constant, produced or consumed gases must be accounted for by a convective flow $\dot{j}_{Q,l}$ between suspension and bubble phase. If the gas is consumed in the suspension phase, the flow $\dot{j}_{Q,l}$ is directed from the bubble phase to the suspension phase. If gas is produced in the suspension phase, a convective flow from the bubbles is assumed.

$$\dot{j}_{Q,l} = \begin{cases} K_q \cdot C_{d,l} & for\ K_q > 0 \\ K_q \cdot C_{b,l} & for\ K_q < 0 \end{cases} \tag{82}$$

The convective exchange rate $K_Q$ is calculated from all heterogeneous reactions rates $r_{g,l}$ based on gases in the suspension phase:

$$K_Q = \frac{R \cdot T}{p} \sum r_{g,l} \tag{83}$$

In this equation, R denotes the universal gas constant, T is the temperature and p the pressure inside the system. The term $C_l \frac{du}{dh}$ in the molar balance equations describes the change of the mass flow of species l with height due to gas velocity changes. In general, the gas flow in the fuel reactor is increasing because either a solid fuel is transferred to the gaseous form, or the gaseous fuel $CH_4$ is reacting, creating two $H_2$ molecules and one CO from one molecule of $CH_4$. Both mechanisms will increase the total flow of gases in the FR. In the AR the oxidation of the OC will decrease the molar gas flow. It is assumed that the gas passes through the suspension phase close to minimum fluidization velocity. This is the reason why only the gas concentrations change in the bubble phase and freeboard due to velocity effects. The chemical reaction in the suspension will lead to differing concentrations inside the bubble and suspension phase. Diffusive mass transfer between the phases is described by the following correlation proposed by Sit and Grace [127] for the gas diffusion resistance $k_G$:

$$k_G = \frac{u_{mf}}{3} + \sqrt{\frac{4 \cdot D \cdot \varepsilon_{mf} \cdot u_b}{\pi \cdot d_v}} \tag{84}$$

In this correlation $u_{mf}$ describes the minimum fluidization velocity, D is the diffusion coefficient, $\varepsilon_{mf}$ stands for the minimum fluidization voidage, $u_b$ is the bubble rise velocity and $d_v$ is defined as the bubble size. The parameter a in equations (79) and (80) describes the ratio of the interfacial area between bubble and suspension phase to the volume of the reactor element under consideration and is calculated with the assumption of spherical bubbles from:

$$a = \frac{6\varepsilon_b}{d_v} \tag{85}$$

### 3.3.3. Calculation procedure for the fluidized bed reactors

In Figure 25, the steps to calculate the module itself are shown. At the start time, the fixed parameters are read into the program. These parameters are not changed during the simulation and, therefore, are only initialized once. These include geometrical values as well as the decision which correlations i.e. for the elutriation of particles, are used. Afterwards, the reactor is discretized into a number of predefined height elements, which are necessary to provide enough intermediate steps for the calculation of fluid mechanics and chemical reactions. Then, for each time step, the operation conditions, such as fluidization velocity and inflows from other process units, are read from the system. Together with the current state of the reactor itself, i.e. its actual bed mass and its conversion state, the solids distribution in the system can be calculated from a fluid mechanic perspective. Knowing the location of the solids, the chemical reaction scheme is necessary to calculate the conversion of gases for each height element.

Fluid mechanics and chemical reactions are iterated as long as gas conversion in plug flow and conversion of perfectly mixed solids match each other. Gas conversion is calculated with an explicit Euler algorithm, whereas the fluid mechanics are calculated with the Sundials DAE solver. When both are iterated, and the calculated values converge to a solution within the predefined tolerances, the outlet flow is calculated along with a new bed mass. The conversion state and particle size distribution of the solids is gathered. The simulation environment Dyssol will decide the time step between two time points automatically in a way to keep the numeric error in the predefined boundaries. Changing inlet and outlet streams are interpolated between the time points.

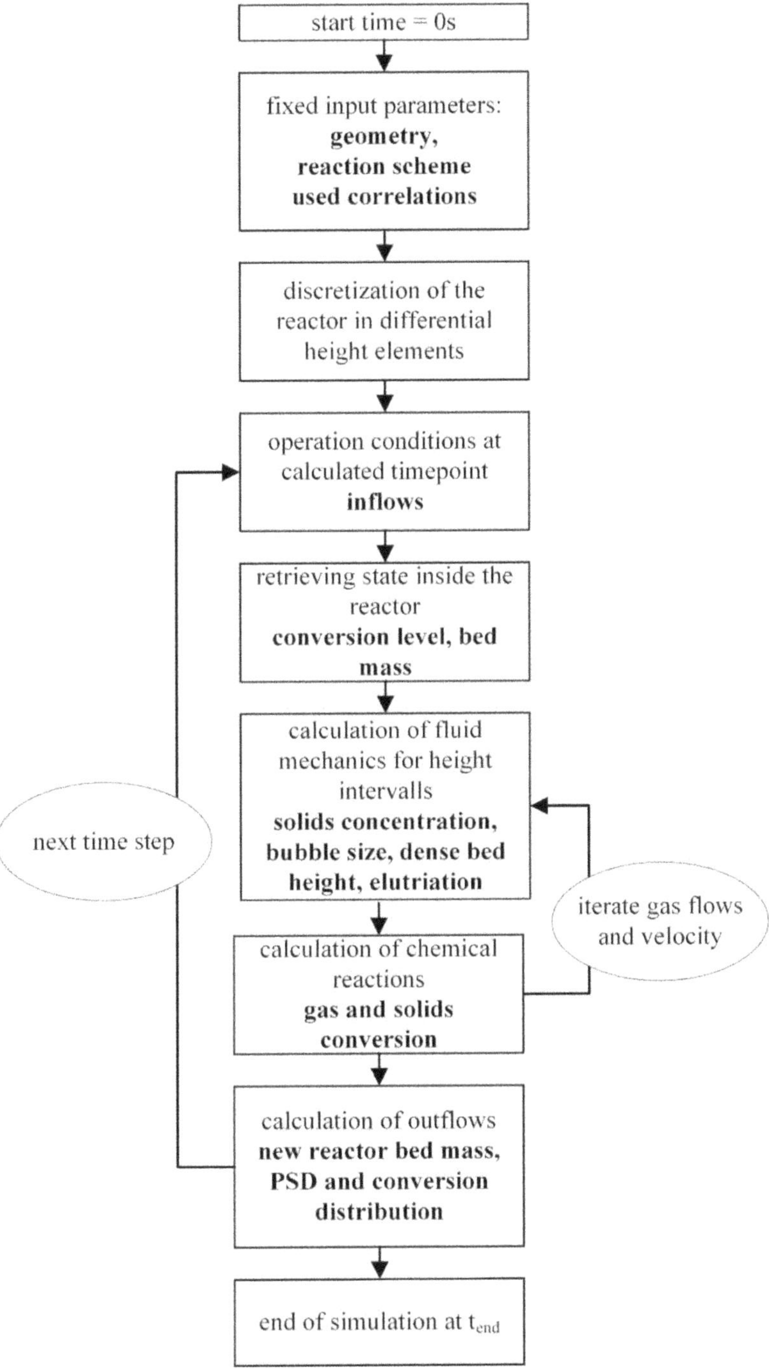


**Figure 25: Calculation procedure for solving the fluidized bed reactor unit.**

### 3.3.4. Modeling of the dynamic behavior for the fluidized bed reactors

For every time point during the calculation the local fluid mechanics are according to the procedure shown in Figure 25. Due to in- and outflows of the reactor, the bed mass can change over time in each respective unit, which has a considerable impact on the local fluid mechanics in the next time point. Besides the changes of masses in the system, other dynamic effects can take place in the reactor system, which must be accounted for. Due to elutriation from the bubbling bed and the not perfect separation of all particles in the cyclone, a steady removal of fine particles from the system takes place. This will lead to a change of the particle size distribution in the reactor system. In the future, also attrition effects can be included, which influence the PSD on a longer timescale.

The most important dynamic effect in the system is obviously the chemical conversion of the solids in the system. Due to the chemical reactions of the fuel and oxygen with the OC, the conversion state in the fluidized bed reactor is changing. Whereas the circulating solids and movement is more a macroscopic effect, the change of the PSD due to extracted fines as well as the conversion state of the OC are taking place on a particle level. This is why both of these effects are represented in so-called distributed secondary particle properties. To ensure that these attributes are transported and manipulated correctly, Dyssol works with transformation matrices. They describe the movement of particles between so-called attribute intervals for each secondary attribute. Skorych et al. [117] have described the core system Dyssol and how transformation matrices transport information between the different process units.

To picture the CLC process the oxidation state of particles and the particle size are described in a conversion distribution and a particle size distribution, respectively. The number of classes in each of the distributions can be set by the user and should be chosen in a way to cover the occurring effects accurately. If only few particle size classes are chosen, for example, then the elutriation effect could be numerically decreased, as the smallest class spans over a too-wide range.

<u>Transient changes of the particle size distribution inside the fluidized bed reactors</u>

The fluid mechanic algorithm calculates the concentration of solids $c_{v,i}$ in each size class i at the top of the reactor H. Together with the total solids concentration at the very same position $c_{v,H}$ one can calculate the volume or mass based size distribution of the exiting particles of both CFB riser reactor and the freeboard of the bubbling fluidized bed:

$$q_3(x_i, H) = \frac{c_{v,i,H}}{c_{v,H} \cdot \Delta x_i} \quad (86)$$

In this equation $\Delta x_i$ is the size of the interval i. The change of cumulative size distribution based on mass/volume $\Delta Q_{3,i}$ becomes for each size class i:

$$\Delta Q_{3,i}(H) = \frac{c_{v,i,H}}{c_{v,H}} \quad (87)$$

The fluid mechanic section provides the bed masses in the reactor at the starting timepoint $m_{reactor,\, t=0}$ and for the next time point $m_{reactor,\, t=1}$. Knowing the distribution and the mass flow of the exiting particles $\dot{m}_{elutriation}$, one can then calculate the distribution of the particles, which remain in the reactor at the next timepoint t = 1:

$$q_3(x_I, \text{reactor}, t = 1) = \frac{q_3(x_i,\text{start}) * m_{reactor,t=0} - q_3(x_i,H) * \dot{m}_{elutriation}(t)}{m_{reactor,t=1} + \dot{m}_{elutriation}(t)} \quad (88)$$

With the inlet and the outlet particle size distributions, Dyssol offers the feature to calculate the needed transformation matrix.

Transient changes of the conversion state inside fluidized bed reactors

A novel mechanism had to be developed to accurately track the conversion of the involved solids inside the system and keep the molar balance between reacted solids and gases. The solids conversion over time $\frac{dX_s}{dt}$ was calculated for a certain timepoint. The particle conversion itself is represented by a conversion distribution $C_{dist}$ similar to a particle size distribution, where a number of conversion classes k between the values 0 and 1 are used to describe the conversion $X_s$. If all particles are fully oxidized, then all particles lie in the class with the value 0. If all particles are fully reduced, then it follows that all particles are in the class with the value of 1, representing full conversion. The reactive particles can then move from one conversion class CC to the next lower one for OC reduction ($\frac{dX_s}{dt} > 0$) or the next higher one for OC oxidation ($\frac{dX_s}{dt} < 0$). The movement itself is carried out with transformation matrices.

The transformation matrix for the reduction reaction is calculated by discretizing the conversion over time and dividing it by the size of the conversion class $\Delta CC_k$, which results in the change of particles in each conversion class k. This resembles a simplified population balance by placing a certain mass of particles in each class and describing their movement between classes over time. Substracting it from 1 gives the remaining particles in the conversion class $U_k$:

$$U_{k,reduction} = 1 - \frac{\frac{dX_s}{dt} * dt}{\Delta CC_k} \quad (89)$$

The transformation matrix for remaining particles in the conversion class takes values (k, k, $U_k$). Transformation matrix movement of the reduced particles to the higher conversion class is described by (k, k + 1, 1 - $U_k$).

The transformation matrix for the oxidation directs the movement of the particles in the other direction:

$$U_{k,oxidation} = 1 + \frac{\frac{dX_s}{dt} * dt}{\Delta CC_k} \quad (90)$$

The transformation matrix for remaining particles in the conversion class: (k, k, $U_k$). Transformation matrix movement of the reduced particles to the lower conversion class is express by: (k, k - 1, 1 - $U_k$).

The transformation matrices for oxidation and reduction reaction are visualized in Figure 26. A change of conversion in the discretized timestep t = 0 to t = 1 of $dX_s = \frac{dX_s}{dt} \cdot dt = 0.04$ is assumed. The size of the conversion $\Delta CC_k$ can be easily seen to be 0.2, meaning that a total of 6 conversion classes are used for the description of the conversion. For the first class with conversion Xs = 0, the oxidation is not calculated, as this class is already fully oxidized. For the reduction reaction, the conversion $X_s = 1$ is not accounted for, as this class is treated as fully reduced. This is expressed by the value of 1 in the transformation matrix at the lower right corner for the reduction and the upper left corner for the oxidation.

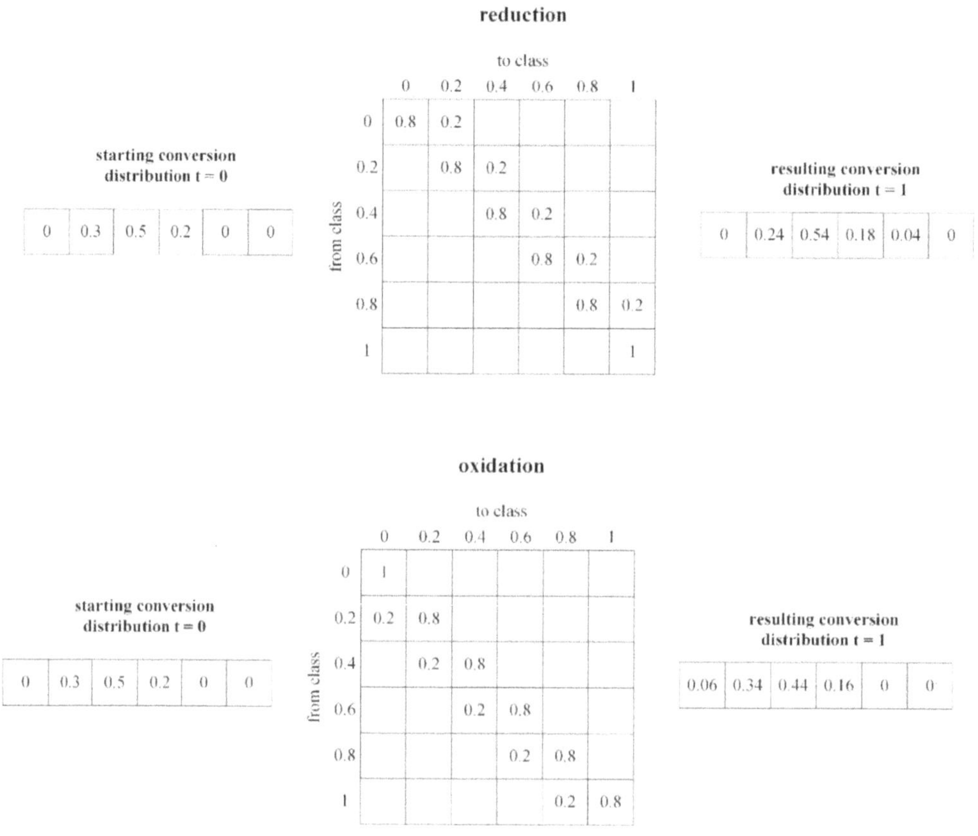


Reduction transformation matrix (from class / to class):

| from \ to | 0 | 0.2 | 0.4 | 0.6 | 0.8 | 1 |
|---|---|---|---|---|---|---|
| 0 | 0.8 | 0.2 | | | | |
| 0.2 | | 0.8 | 0.2 | | | |
| 0.4 | | | 0.8 | 0.2 | | |
| 0.6 | | | | 0.8 | 0.2 | |
| 0.8 | | | | | 0.8 | 0.2 |
| 1 | | | | | | 1 |

Oxidation transformation matrix (from class / to class):

| from \ to | 0 | 0.2 | 0.4 | 0.6 | 0.8 | 1 |
|---|---|---|---|---|---|---|
| 0 | 1 | | | | | |
| 0.2 | 0.2 | 0.8 | | | | |
| 0.4 | | 0.2 | 0.8 | | | |
| 0.6 | | | 0.2 | 0.8 | | |
| 0.8 | | | | 0.2 | 0.8 | |
| 1 | | | | | 0.2 | 0.8 |

**Figure 26: Reduction and oxidation changes to the conversion distribution. Arbitrary values: $\frac{dX_s}{dt} \cdot dt = 0.04$, $CC_k = 0.2$.**

If two material streams or a material stream and a solid holdup with differing conversion distributions are mixed, a novel distribution is calculated by simply adding all particles in the same class together. Then it is calculated how many of the particles are in the respective class according to a mass balance. In the current state of the modeling work, an overall conversion $X_s$ is calculated from all classes k in the conversion distribution $C_{dist}$ according to equation(91).

$$X_s = \sum_0^k C_{dist}(k) \cdot k \cdot \Delta CC_k \tag{91}$$

The reaction mechanism described before will use the overall conversion $X_s$ and not calculate the conversion rate for each conversion class separately. This was done, because the conversion rate $\frac{dXs}{dt}(X_s)$ was usually determined assuming a non-distributed value for $X_s$. In the future one could think of calculating the reaction rates based on the distributed conversion. Further, the reaction rates are not connected to the size of the particles, for the same reason. The rate determining parameters $E_a$ and $k_0$ were usually determined for a defined particle collective with a non-uniform size distribution and not for uniform particle sizes.

### 3.3.5. Cyclone unit

For the cyclone a semi-empirical model by Muschelknautz and Trefz [128], which has proven to be applicable in a broad range of operation conditions, was used. This model implies a separation of particles and gas by two different mechanisms depicted in Figure 27. The first one is wall separation due to exceeding the saturation carrying capacity, which causes the formation of a strand directly leading down the wall to the solids exit. The second is the separation inside the inner vortex. A minor part of the inlet stream is moving directly to the upper outlet without being separated at all, the so-called overflow.

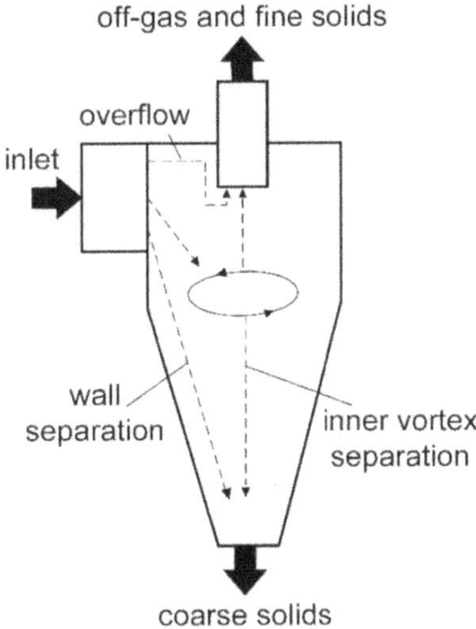


**Figure 27: Cyclone model with separation mechanisms. Adapted from Muschelknautz and Trefz [128].**

Wall separation occurs when the inlet solids loading $\mu$ exceeds a certain threshold value $\mu_G$. Solid loads above this value are directly separated. Because preferably bigger particles will be separated at the wall, a classification effect will cause a finer particle size distribution for the particles in the inner vortex. For flows with high solids loading, this is the predominant separation mechanism. Own calculations show that in CFBs the flow usually exceeds $\mu_G$ by orders of magnitude.

The separation in the inner vortex decides which particles will ultimately leave the system either with the overflow or with the underflow. Inside the inner vortex a cut size $d^*$ is defined, which decides which fraction in the particle size class d is separated in the inner vortex according to the separation efficiency function $\eta_f(d)$ in equations 13-15. The parameter D varies for certain cyclones between 2 and 4 and is mostly set to the value of 3.

$$\eta_f(d) = 0 \qquad \text{for } d < \frac{d^*}{D} \tag{92}$$

$$\eta_f(d) = 0{,}5 * \left\{1 + \cos\left[\pi\left(1 - \frac{\log\frac{d}{d^*} + \log D}{2\log D}\right)\right]\right\} \qquad \text{for } \frac{d^*}{D} < d < D \cdot d^* \tag{93}$$

$$\eta_f(d) = 1 \qquad \text{for } D \cdot d^* < d \tag{94}$$

This model was recently shown to have shortcomings concerning the population balances and the separation of a solids mixture with different densities of the solids [129]. The model assumes a fictitious PSD for the inner feed for the vortex separation. Additionally, the multiphase flow is separated by means of particle size considering only a single particle density, which becomes problematic when a mixture of different solids has to be separated. This is why Redemann et al. [129] proposed a model relying on the terminal velocities of particles instead of their diameters. The separation efficiency curve $T(u_{t,i})$, which determines the part of solids that can be separated in the inner vortex, is hence dependent on the terminal velocity $u_{t,i}$ for each defined terminal velocity class i. The value $u_{t,50,e}$ describes the terminal velocity $u_t$ of a representative particle of a flow, with 50% of the particle mass belonging to particles with terminal velocities below this value. Which share of the flow is present inside the inner vortex is expressed by the share ratio a.

$$T(u_{t,i}) = (1 - a) \cdot T^*(u_{t,i}) + a \tag{95}$$

$$T^*(u_{t,i}) = \frac{1}{1 - \sqrt{\frac{u_{t,50,e}}{u_{t,i}}}\exp\left\{a\left[1 - \sqrt{\frac{u_{t,i}}{u_{t,50,e}}}^{3}\right]\right\}} \tag{96}$$

The flowsheet environment Dyssol allows the implementation and easy change of these different sub models. Hence different sub models can be investigated for their applicability to the respective modeling work.

It should be noted here that the cyclone is modelled with a steady state process model. This is done because of the small particle hold-up which is to be expected inside the unit. Hence, a particle laden flow entering the cyclone is instantaneously separated into the gas flow with small entrained particles leaving the system via the overflow and a coarse particle underflow circulating back.

### 3.3.6. The loop seal unit

For the dynamic simulation of the CLC reactor system, the modeling of the loop seals is important, as it will act as a hold-up for particulate material. The mass, $m_{LS}$, which is stored in the loop seal, is strongly depending on the pressure environment around the loop seal and, with it, from the FR and AR. Figure 28 shows the basic geometry of a loop seal as it is used in the experimental plant and, therefore, is the geometry that was modeled into the Dyssol environment. Solids enter the siphon in the supply chamber (SC) pass through the slit and pile up in the recycle chamber (RC). In the end, they pass over the weir and return back to the respective fluidized bed reactor. A loop seal is modeled including the standpipe such that the solids in the standpipe are included in the total mass of the SC. The total mass $m_{LS}$ inside the loop seal at any time point is the sum of masses in the recycle chamber $m_{RC}$ and the supply chamber $m_{SC}$:

$$m_{LS} = m_{SC} + m_{RC} \tag{97}$$

A loop seal similar to the one operated in Hamburg was investigated by Basu and Cheng [130–132], which describes the movement of solids through the loop seal. In their approach, which was integrated into the present flowsheeting environment, the solid masses distribute according to a pressure balance. In contrast to their findings, it was visible in Hamburg's cold facility that depending on the pressure distribution and gas velocities, also a fluidized bed can form in both the SC and the RC and not only in the RC. Depending on the relative velocity between gas and solids in the SC, the fluid mechanic models for a fluidized bed from the section above or for a moving bed from Basu and Cheng are applied.

Several factors determine the mass of solids inside such a loop seal. Similar to the fluidized bed reactors, solids will flow out of the loop seal when the height of the bed in the RC is higher than the weir. The main determining factor for the height of the both chambers is the pressure difference:

$$\Delta p_{LS} = p4 - p1 \tag{98}$$

between the FR p4 side and the AR p1 side determines the height difference of solids in both chambers. By adjusting the height of solids in the standpipe, the loop seal will balance out the two reactors. The balancing effect of the loop seal is limited by three factors. Firstly, the solids always have to reach the slit height to prevent short-circuit of the gas. Secondly, the height between slit and the overflow defines the highest possible height of solids in the recycle chamber RC. Lastly, the total height of the supply chamber SC and the standpipe above restrict the maximum amount of solids in it. The siphon model was implemented to cover all described effects and for every time instantly the bed height difference for both chambers is calculated via a force balance.

The pressure difference between the two sides of the loop seal siphons is usually determined by other equipment, the cyclone with its pressure drop and the AR on the one side and the FR on the other side. The accurate determination of this pressure drop needs a backward calculation sequence for the pressure. Because of the used bubbling bed model, the solids volume concentration $c_v$ of the bed in each chamber changes, depending on the height h above the distributor of the loop seal siphon. Currently, the pressure difference between both sides of the loops seal is set constant. In the future, this fixed pressure difference over the loop seal can be replaced with time resolved pressure drops from FR and AR simulations.

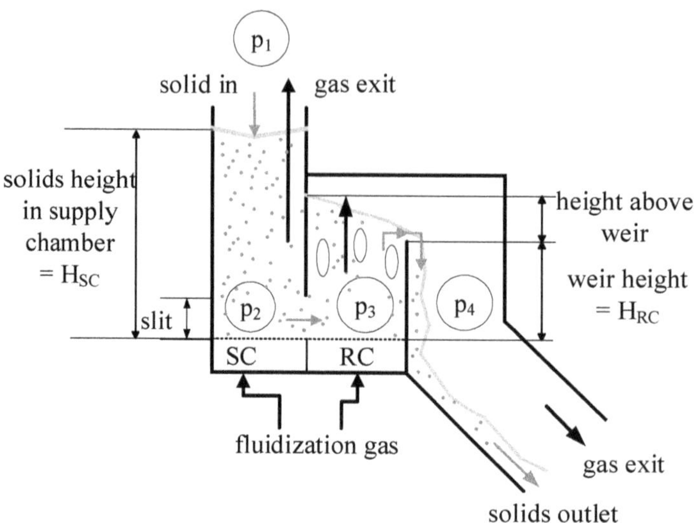


**Figure 28: Sketch of siphon loop seal S2 as operated at Hamburg University of Technology. SC and RC denote the supply and the recycling chamber, respectively.**

Bareschino et al. [133] showed that the solid mass flux is not influenced by the fluidization velocity inside the loop seal, but that the velocity inside the riser reactor has the major influence on solids circulation. Furthermore, they showed that basically no gas flux from the recycle chamber towards the supply chamber SC occurs, whereas a significant amount of supply chamber gas leaves through the recycle chamber RC outflow. This is in accordance with the findings by Thon [15], who traced gas leakage in the same direction. Because of this effect the model assumes a total outflow of fluidization gas with the overflow.

The loop seals are implemented into the system to allow for dynamic changes of the bed mass and will arrange the pressure drop difference of the two chambers:

$$\Delta p_{LS} = \int_0^{H_{SC}} \rho_s c_v(h)\, g\, dh + \int_0^{H_{RC}} \rho_s c_v(h)\, g\, dh \tag{99}$$

The time dependent bed mass is then given by the mass in and outflows in Figure 28. The outflow out of the loop seal is calculated just as it is done in the bubbling fluidized bed reactor with equation (100).

$$\frac{dm_{LS}}{dt} = \dot{m}_{LS,in} - \dot{m}_{LS,out} \tag{100}$$

# 4. Results and Discussion

## 4.1. Operation of the CLC pilot plant with different fuels

### 4.1.1. Gas concentration measurements at AR and FR

Fine lignite injected on lower stage

A typical operation with the finer fraction of the lignite resulted in the gas concentration measurements of Figure 29. During the whole course of the operation, noticeable $H_2$ and CO concentrations above 1 vol.% were detected at the outlet of the FR system. Also, a small fraction of $CH_4$ is visible in the experimental data. This can be traced back to the unfinished conversion after the first stage, where gas concentrations of 3 vol.% $CH_4$, 10 vol.% CO and 15 vol.% $H_2$ can be detected. Comparing the conversion after the first and the second stage one can see the significant effect of the second stage of the FR system. Analyzing the differences between the conversion of a coarse and a fine fraction of lignite one can see that a much higher amount of combustible gases is visible after the first stage of the FR system. This can be traced back to gasification and devolatilization, which for the fine fraction appears to take place more in the upper part of the bed or in the freeboard. The coarse fraction seems to convert mainly in the dense bubbling bed, where most of the combustible components are oxidized completely.

Gas concentrations of the AR for the fine fraction show a slightly smaller amount of $CO_2$, even though the fuel input was higher. The $O_2$ concentration demonstrates the stable and constant operation of the system over more than three hours of fuel supply.

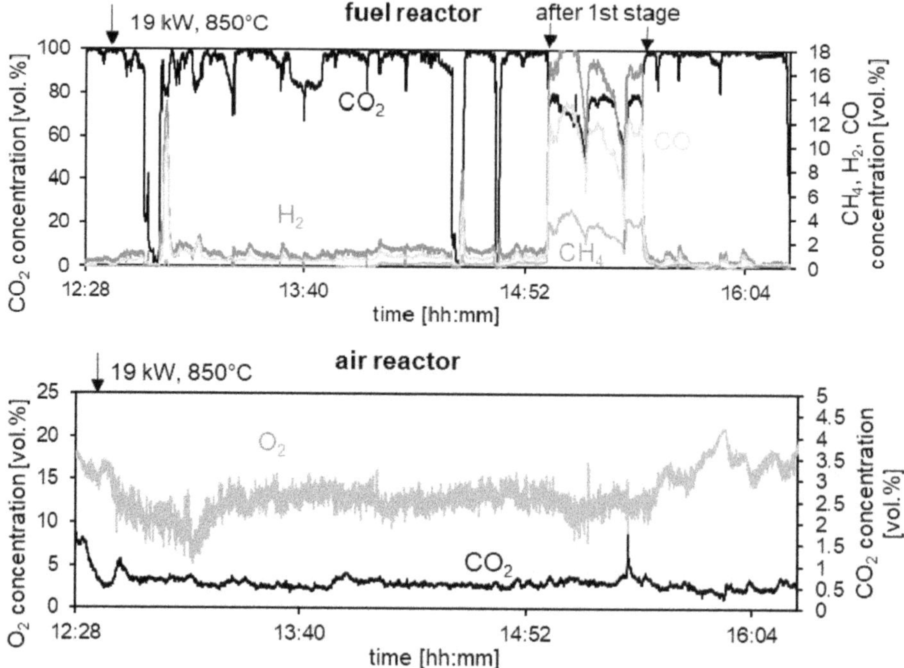


**Figure 29: Gas concentration measurements for fine lignite at 850°C injected into the lower stage.**

<u>Fine lignite injected into the upper stage</u>

Lignite dust was injected into the upper stage of the FR system and the gas concentration measurements are shown in Figure 30. The high concentrations of around 4 vol.% $H_2$ and 2 vol.% CO at very low solid fuel input indicate the bad conversion if the lignite dust is injected into the upper stage of the system. A higher temperature in the reactor system increased these two concentrations even more. The absence of $CH_4$ could be explained either by reaction of CH4 to $H_2$ and CO, which might be favored, or a reaction of $CH_4$ with steam again towards $H_2$ and CO. The injection of fine lignite into the upper stage of the FR led to a total avoidance of carbon slip to the AR, which can be read from the $CO_2$ measurements at 0 vol.% at the AR exit.

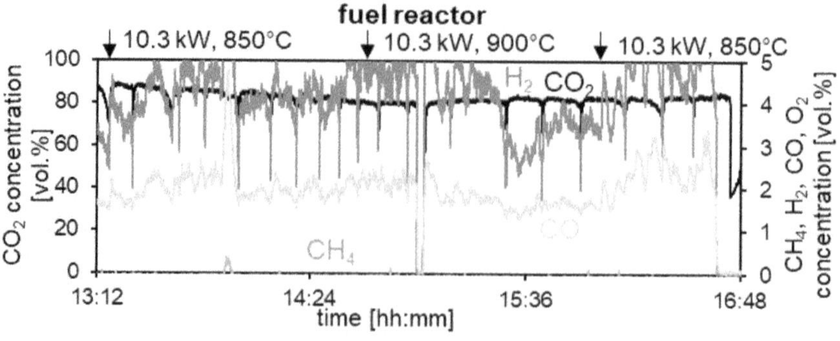

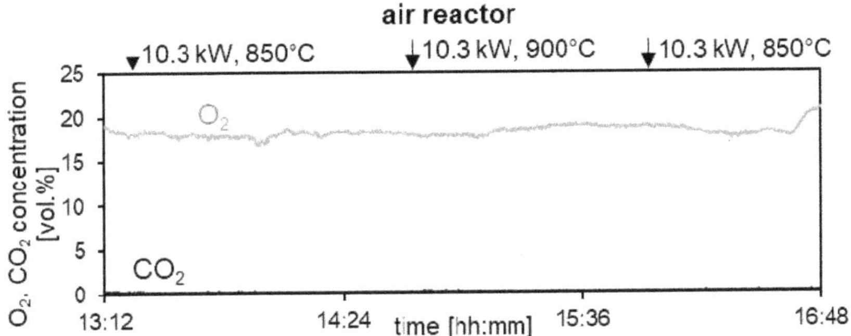


**Figure 30: Gas concentration measurements for fine lignite at 850°C and 900°C injected into the upper stage.**

<u>Coarse lignite injected on lower stage</u>

Coarse lignite coal was fed into the system at the lower stage of the FR at 850 °C and 900 °C. Due to the screw feeding system the power input changed from 10 $kW_{th}$ to 17 $kW_{th}$ during the temperature increase. Figure 31 shows the volumetric gas concentrations of the FR and the AR. During the experiment, the fluidization velocities were kept constant. The concentration of oxygen in the air reactor decreased with a short time delay, which is due to the solid hold-up inside the system. $CO_2$ in the AR rose due to the increase fuel load, but the heating to 900°C lowered it again to around 1 vol.%.

Gas concentration measurements taken from the sampling port above the first stage of the FR system reveals that some combustible $H_2$ and CO at around 1 vol.% is detectable, which is oxidized in the second stage. After the second stage, only around 0.2 vol.% CO and no $H_2$ is measured.

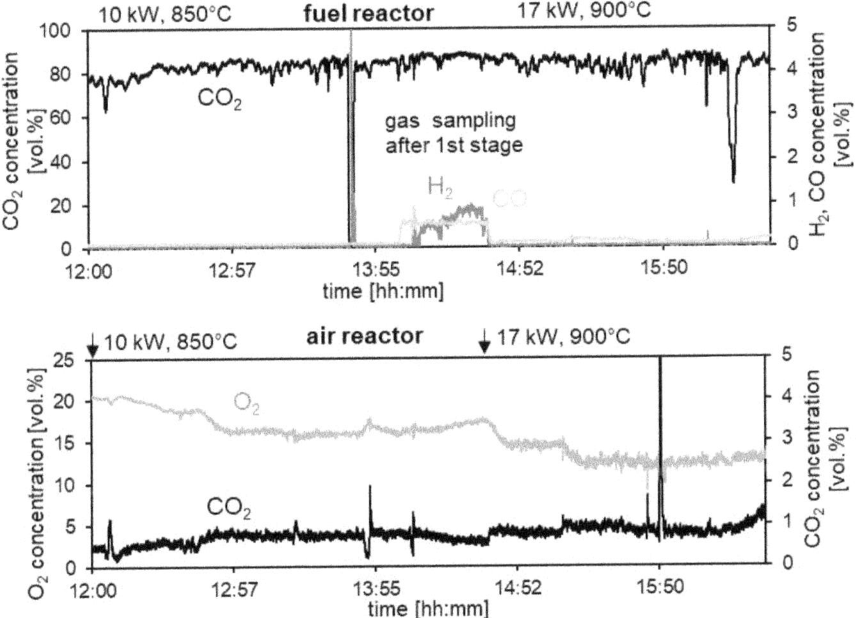

**Figure 31: Gas concentration measurements for coarse lignite at 850°C-900°C.**

Coarse lignite injected into the upper stage

Due to the unpromising results found for the fine lignite dust injected into the upper stage, the injection on the upper stage for coarse lignite was not carried out. It was expected to create a higher flow of volatiles leaving the reactor unconverted and only decreasing the carbon slip minimally.

Fine bituminous coal injected into the lower stage

Injecting fine bituminous coal into the lower stage of the FR system resulted in the gas concentration measurements depicted Figure 32. With this size fraction, some $H_2$ and CO were encountered in the FR off-gas. In the AR, around 3.4 vol.% $CO_2$ is visible. Compared to both size fractions of lignite, the amount of $CO_2$ in the AR off-gas was much higher, which results in a significant worsening of $CO_2$ separation ability with bituminous coal.

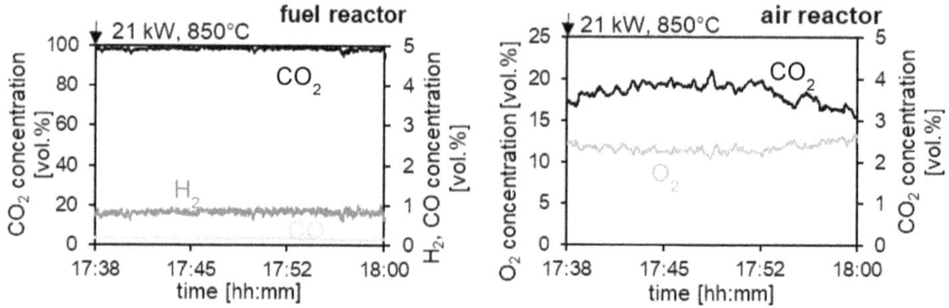


**Figure 32: Gas concentration measurements for fine bituminous coal injected into the lower stage at 850°C.**

Fine bituminous coal injected into the upper stage

Gas concentration measurements during the conversion of fine bituminous coal injected in the upper stage are depicted in Figure 33. The bituminous coal was injected into the upper stage of the FR system to check, whether the carbon slip to the AR could be decreased by giving the char the time to convert first in the upper stage and then in the lower stage before being dragged over to the AR. From the FR gas concentrations, one can see that an increase in fuel mass flow will generate more CO in the off-gas. When fuel power was increased to 26.7 $kW_{th}$, even methane and hydrogen were visible in considerable amounts. Comparing the upper and lower stage injection regarding the carbon slip, the situation could not be improved. During upper stage injection at 17.8 $kW_{th}$ around 3.5 vol.% $CO_2$ were traced in the AR compared to 3.4 vol.% $CO_2$ during the injection into the lower stage with a fuel power of 21 $kW_{th}$. This constitutes even to a worsening in carbon capture, when fine bituminous coal is injected into the upper stage. A possible explanation could be the fluid mechanic situation in the FR reactor system, where the upper injection might not mix the fine bituminous coal sufficiently into the bed. Another explanation could lie in the placement of the standpipes, which might give the fine bituminous coal char in the upper stage a shorter way to travel from the injection point through the standpipes towards the AR.

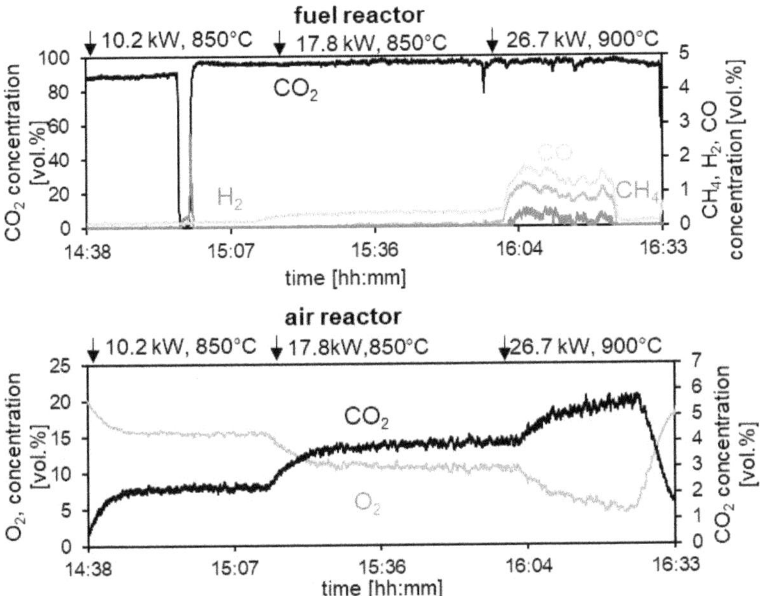


**Figure 33: Gas concentration measurements for fine bituminous coal at 850°C into the upper stage with varied fuel mass flows.**

<u>Coarse bituminous coal injected into the lower stage</u>

Changing the fuel type to coarse bituminous coal had a significant effect on the conversion behavior plotted in Figure 34. In the FR system minor combustible gases are exhausted, which is most probably due to the lower volatile content and the slower gasification compared to lignite. A recognizable amount of 6.5 vol.% of $CO_2$ is detected in the AR, much higher than the ones seen with fine bituminous coal. The slow gasification of bituminous coal appears to cause a remarkable carbon slip from the FR system to the AR. Increasing the temperature in the system from 850°C to 900°C lowered the $CO_2$ concentration in the AR to 5.7 vol.%, meaning that bigger shares of the fuel are converted in the FR system.

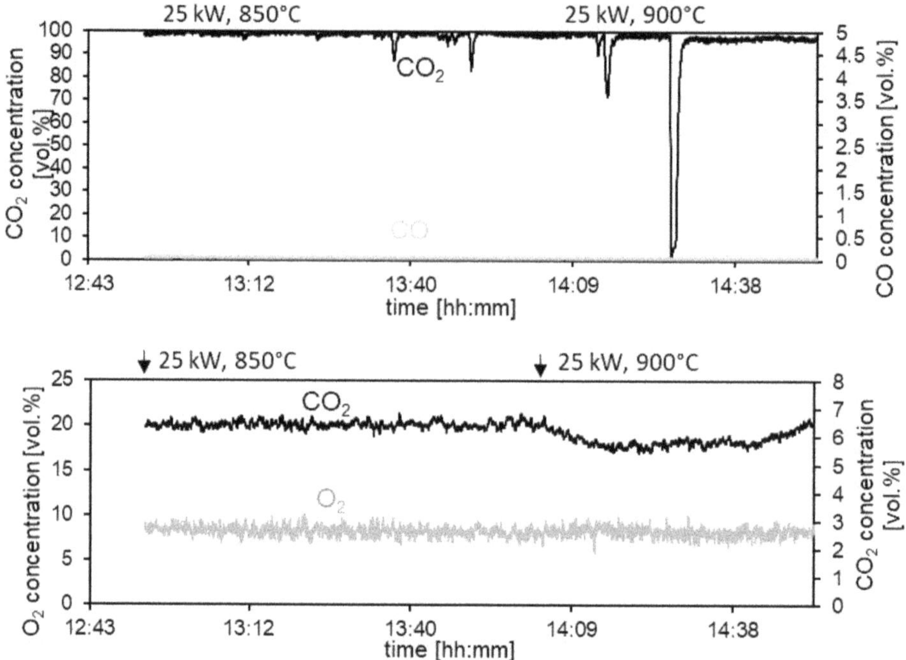


**Figure 34: Gas concentration measurements for coarse bituminous coal injected into the lower stage at 850°C and 900°C.**

Coarse bituminous coal injected into the upper stage

The injection of the coarse fraction of the used bituminous coal into the upper stage was intended to decrease the carbon slip from the FR to the AR. This would improve the relatively poor carbon capture when using the fuel seen in the evaluation above. The char was expected to be gasified in the upper stage partly, before being dragged to the lower stage via the standpipe. There, further char gasification could take place. Figure 35 shows that the carbon concentration in the AR is around 3.5 vol.% at 800°C and around 3.2 vol.% at 900°C for upper stage injection. For the lower stage injection values of around 6 – 7 vol.% during similar fuel power could be seen. This constitutes a remarkable reduction in $CO_2$, which cannot be separated and, hence, improving the plant performance. With the coarse bituminous coal injected into the upper stage, the amount of unconverted gases is not increasing much, meaning that the oxygen demand will not increase significantly.

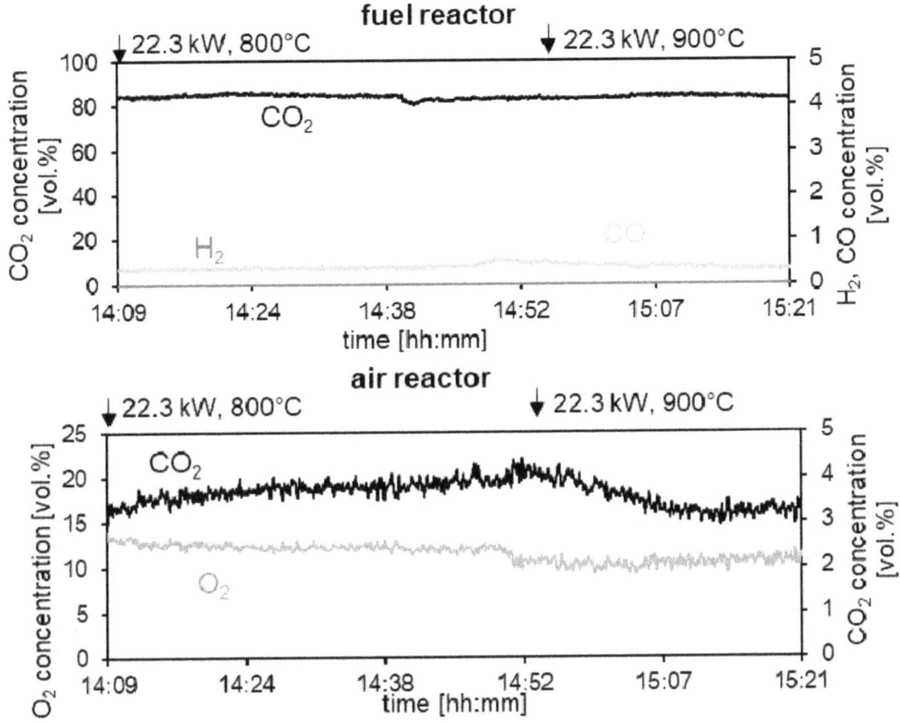


**Figure 35: Gas concentration measurements for coarse bituminous coal injected into the upper stage at 850°C and 900°C.**

<u>Methane injected along with fluidization gas</u>

A gaseous fuel, methane, was used to examine the implication the fuel choice has on the CLC process. Experimental results from the methane run are plotted in Figure 36. After the first stage, still 35 vol.% of methane can be detected in the off-gas, where the inlet concentration was 62 vol.%. After the second stage the methane outlet concentration decreased to 5 vol.%. The $H_2$ concentration was at its maximum 1 vol.% after the first stage and the CO was detected at 0.5 vol.% there. In the exhaust stream both gas concentrations were lower. Concluding this experiment, one can say that the use of methane resulted in the highest concentrations of combustible gases by far, especially considering the relatively low fuel input of 15 kW. The AR concentration measurements show around 0.2 vol.% of $CO_2$, which stems from leakage from the FR to the AR via both loop seals. The oxygen concentration was stable at 15 vol.% showing a stable and steady state operation during this experimental run.

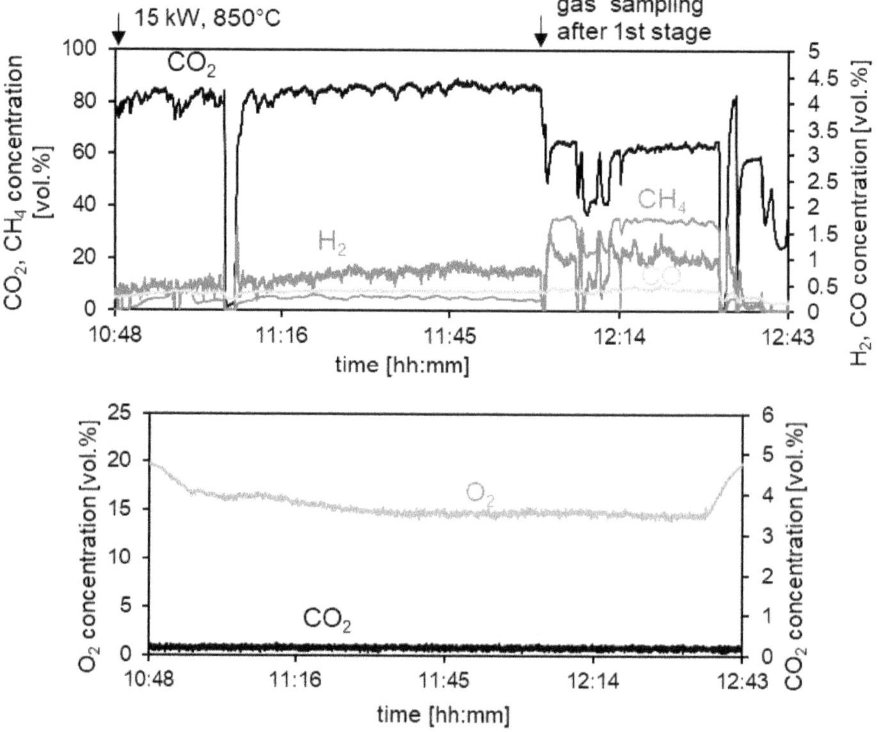


**Figure 36: Gas concentration measurements for methane at 850°C. At 12:00 gas extraction was switched to the first stage in the FR shortly before the inner gas distributor of the second stage.**

Coarse Biomass conversion during injection into the lower stage

The gas volumetric concentrations of FR and AR exhaust gas were measured during the injection of biomass and are shown in the Figure 37. One can see that the $H_2$ and CO concentration were around 0.8% and 0.7%, respectively at the outlet of the FR, while the $CH_4$ fraction is almost invisible. $CO_2$ concentrations up to 82% were obtained. It means that the carbonaceous gases were converted to a high degree through the reaction with the OC. The remaining percentages is due to the leakage from AR, namely $O_2$ and $N_2$. Previous experiments at this pilot plant revealed that at the lower loop seal S2, gas from the FR leaks to the AR [9]. In the upper loop seal S1 however, gas from AR leaks to FR, which consequently leads to a dilution of $CO_2$.

The dry volumetric gas concentrations from the upper part of the first stage were also measured and can be seen in Figure 37.As it shows there, concentrations of $CH_4$, CO and $H_2$ of 0.9%, 2.1 and 2.5%, respectively are detected, which are significantly higher than after the second stage. This result demonstrates that the conversion in the first stage was not as much complete and relatively high amounts of combustibles are present before the second stage. One has to keep in mind here that around 70% of the $CO_2$ originates from the fluidization and injection gases, which reduces the volumetric concentration from the gases stemming from fuel conversion.

The dry volumetric concentrations of $O_2$ and $CO_2$ in the AR were measured at the outlet of it. During the experiment, the gas concentrations are quite stable, which demonstrates that a steady operation is achieved during the experiment. The relatively high value of $O_2$ there shows that a higher fuel injection rate would have been possible from the oxygen supply point of view. However, the injection pipe with a diameter of 8 mm on the one side and the low density of the biomass on the other side were limiting the fuel input.

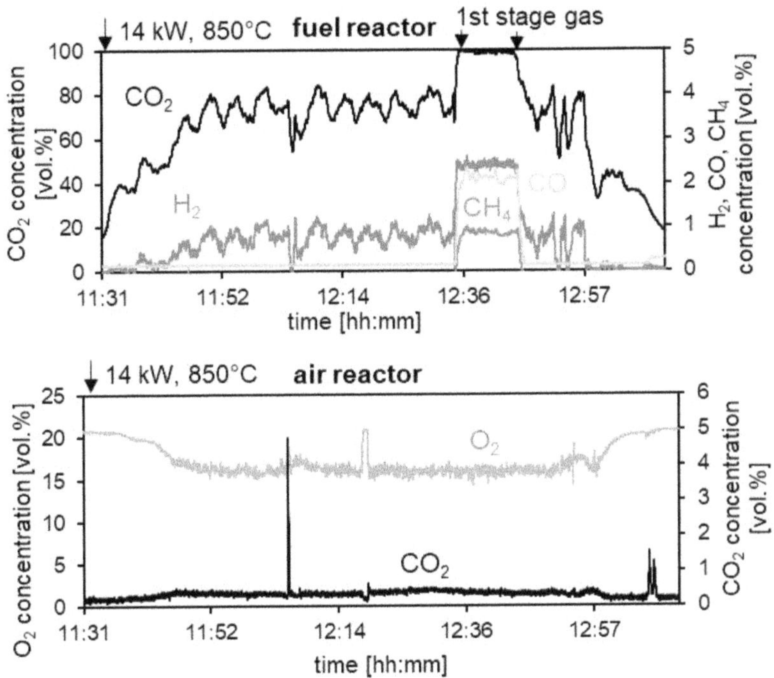


**Figure 37: Gas concentration measurements for biomass conversion at 850°C. Extraction of gases from the first stage from 12:34-12:44.**

### 4.1.2. Performance parameters for the pilot plant operation

<u>Carbon capture efficiency/oxide oxygen efficiency</u>

The carbon capture and fuel conversion efficiencies are plotted in Figure 38. Big deviations could be found between the carbon capture during the combustion of lignite and bituminous coal. The carbon capture was close to 100% for lignite dust, but the bituminous coal showed lower capture rates of 50–62% for the fine fraction and 12–47% for coarse particles. The reason for this is the notably lower conversion rate of the bituminous coal char compared to the lignite. This causes a higher concentration of char in the bed of the lower stage of the FR and thus a higher amount of char, which flows together with the circulating OC to the AR, in which it is combusted completely. A higher fuel flow in experiment 4 than in 5 decreases the capture efficiency. An increased temperature, that is, in experiments 2, 6, 7, and 9, leads to a better conversion and hence better capture in the FR. For bigger fuel particles, the temperature seems to play an even more substantial role for carbon capture.

From the AR molar balance, the oxide-oxygen capture efficiency $\eta_{OO}$ was calculated for all experiments and is plotted as well in Figure 38. The efficiency gathered from the AR gives slightly higher values of carbon capture. Both balancing methods for the AR and FR, respectively, have minor weaknesses with regard to the current measurement approach at the facility. On the AR side, a small flow of purge air is added to keep the pressure-measurement pipes free of solids, which could explain the slightly higher capture values. However, fuel reactor balancing is dependent on the evaluation of the dust after the experimental day and the manually recorded coal injection, which is not totally precise. If both carbon capture parameters are closely aligned, one can argue that both AR and FR balancing is accurate. Both performance indicators together provide a good picture of the carbon capture potential of the plant.

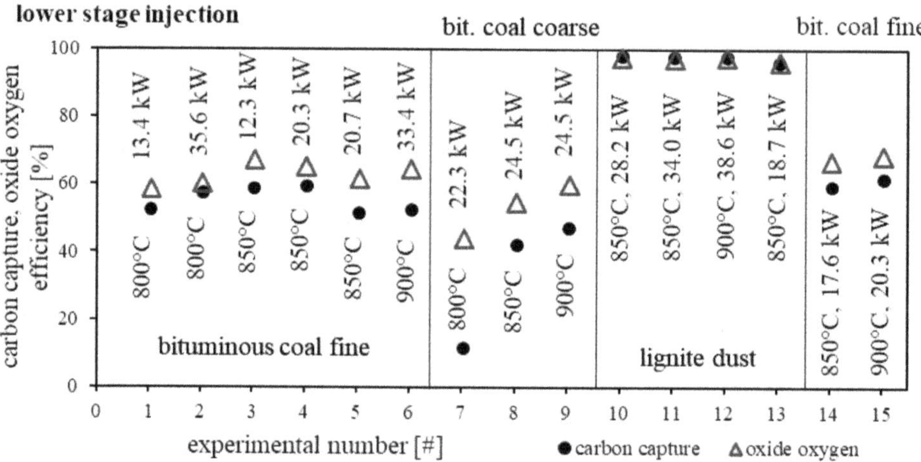


**Figure 38: Carbon capture (•) and oxide oxygen efficiency (Δ) for the experimental campaigns on the lower stage injection of lignite dust and coarse and fine bituminous coal.**

Poor carbon capture efficiencies of a maximum of $\eta_{OO}$ of 69.3% and $\eta_{CC}$ of 62.7% for fine bituminous coal during injection into the lower stage were detected. The rates were considerably lower for the coarse coal. This was the reason, why it was tried to inject the bituminous coal into the upper stage of the FR instead of the lower stage. This was done to use the two-stage design in a way to give the char more time before it will slip to the AR. First the char has the time to gasify in the upper stage, before it will be dragged to the lower stage eventually. In the lower stage, it will have time to gasify again. The experimental results are shown in Figure 39, where one can see that #16 with fine bituminous coal saw an outstanding increase in carbon capture to a maximum of $\eta_{OO}$ = 87.8 and $\eta_{CC}$ = 76.8, respectively. Remarkable increases were also found in the experiments with coarse bituminous coal in #17-#19 even at the low temperature of 800°C. The experimental campaign dedicated to the dynamics during operation (#20 - #22), showed somewhat lower carbon capture during upper stage injection, but still considerably higher than before. Carbon capture during the upper stage injection of lignite dust (#23 - #25) could not be improved as carbon capture were already close to completion during the experiments with lower stage injections (> 96.6%).

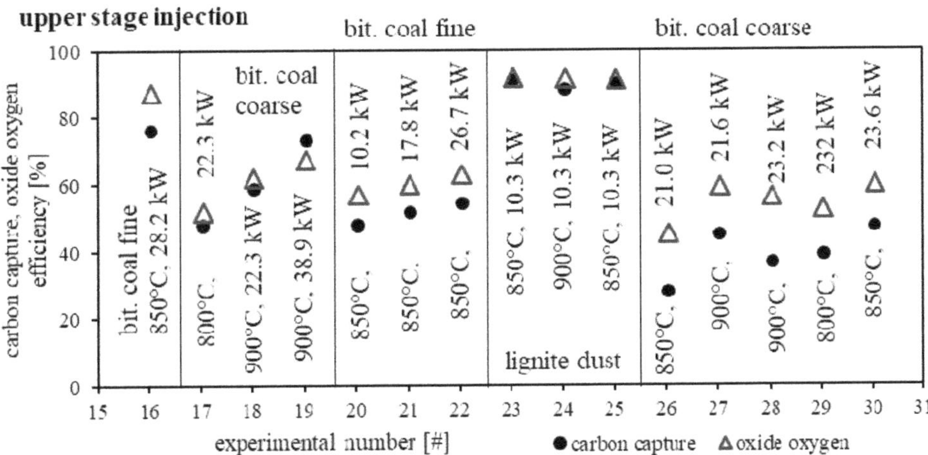


**Figure 39: Carbon capture (•) and oxide oxygen efficiency (Δ) for the experimental campaigns on the upper stage injection of lignite dust and coarse and fine bituminous coal.**

In a next step coarse lignite coal was injected into the lower stage to see the influence of the particle size on the conversion behavior. Coarse lignite showed reasonable capture rates constantly above $\eta_{CC}$ and $\eta_{OO}$ above 75% as indicated in Figure 40. Compared to finer lignite dust, the rates were lower, which indicates that the bigger fuel particles need a longer timeframe to gasify and, hence, are more likely to be carried over to the AR, where the carbon cannot be captured. Methane capture rates were over 95% meaning that around 5% of the carbonaceous gases slip from the FR to the AR. Experiment #38 was conducted to investigate dynamic behavior of the unit but could well reproduce the results in earlier campaigns. Biomass showed really good carbon capture behavior in the system, which can be traced to the really low fraction of fixed carbon in this fuel. This low amount of carbon is even more easily gasified compared to higher ranked coals.

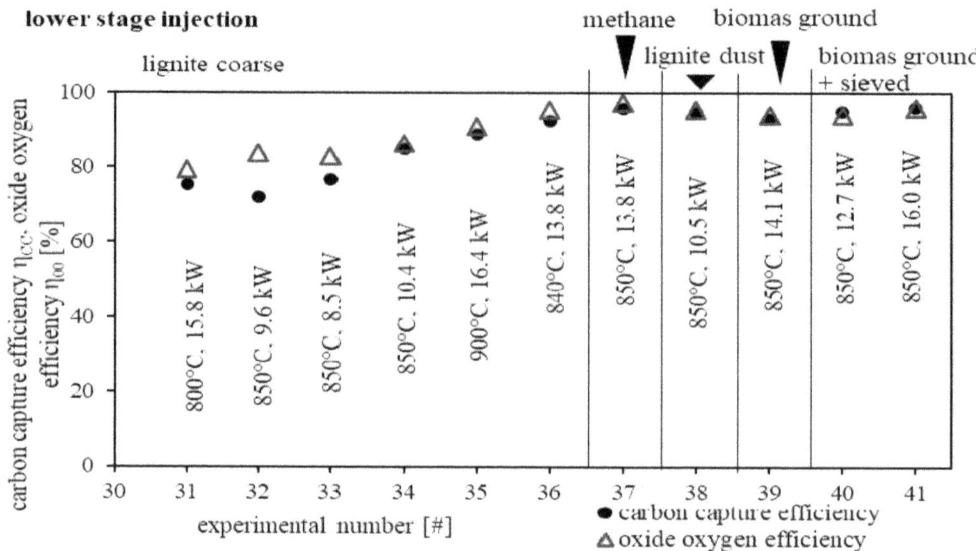


**Figure 40: Carbon capture (●) and oxide oxygen efficiency (Δ) for the experimental campaigns on the lower stage injection for coarse and fine lignite as well as methane and two fractions of biomass.**

The overall fuel conversion is above 95% throughout all experiments, which is demonstrated in Figure 41. Lowest values were found for coarse bituminous coal at 800°C in the lower stage ($X_c$ = 95.4%) and the upper stage injection of fine lignite ($X_c$ = 95.0%). The upper stage injection of fuels did not decrease the solid fuel conversion on a noticeable scale compared to lower stage injection. In the pilot plants of other institutes, summarized in Table 2, the solid fuel conversion was similarly high. Generally, it can be said that the solid fuel conversion is not a major issue also in this reactor setting.

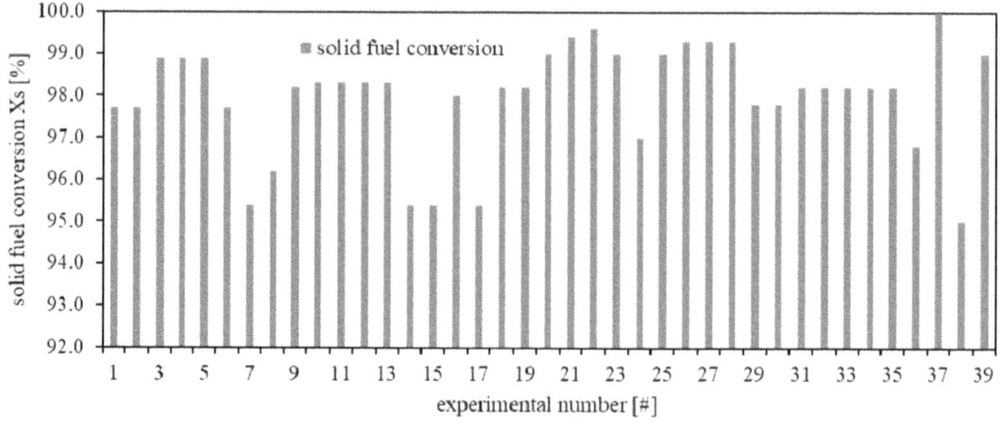


**Figure 41: Solid fuel conversion in the CLC pilot plants for all investigated fuels.**

Low concentrations of combustible gases were detected in the off-gas of the second stage of the FR, which was shown in the section about the gas concentration measurements. This led to low values of the oxygen demand of 0.1 – 1.8% shown in Figure 42. If lignite was injected, more combustible gases leave with the FR off-gas, which leads to a slightly higher oxygen demand. However, the $CO_2$ concentration at the AR outlet for lignite is lower than that of bituminous coal. This shows that a greater amount of lignite is converted inside the FR. The extremely low value for the oxygen demand for experiment #7 with coarse bituminous coal is coupled with a low conversion of coal in the FR in general. Looking at the absolute values of the oxygen demand, which are all below 2%, it can be stated that the present reactor system converts the gases from the fuel gasification for both fuels successfully. This is mainly because of the effect of the FR design as other plants with only a single stage, summarized in chapter 2.2.3, tend to have significantly higher oxygen demands of around 20%.

**lower stage injection**

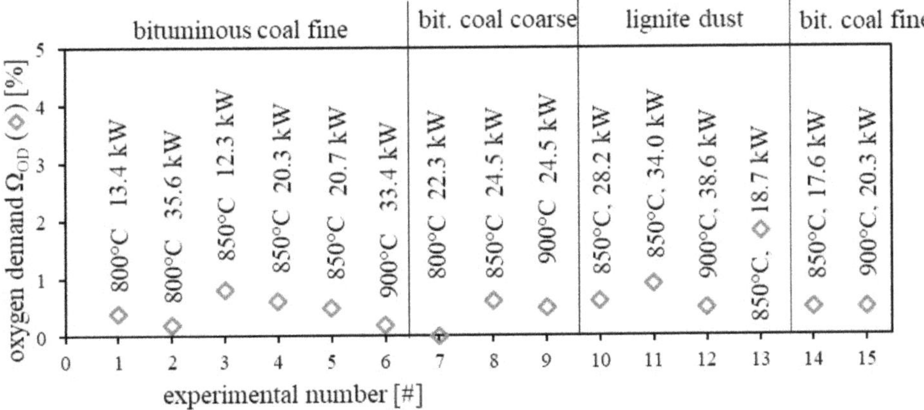


**Figure 42: Oxygen demand experienced during the experimental campaigns, when solid fuel was injected into the lower stage.**

The upper stage injection of the bituminous coal resulted in similarly low concentrations of combustibles in the off-gas as the lower stage injection. This lead again to an oxygen demand of less than 1% for this coal shown in Figure 43. Lignite dust on the other hand showed considerably higher concentrations of $CH_4$, CO and $H_2$ in the exhaust and, with it, the oxygen demand was constantly over 10% already at low fuel mass flows. From these results one can follow that the upper stage injection is favorable for fuels with a high amount of char and little volatiles. High volatile fuels lose the second stage for the gas conversion and were proven to be well converted during lower stage injection.

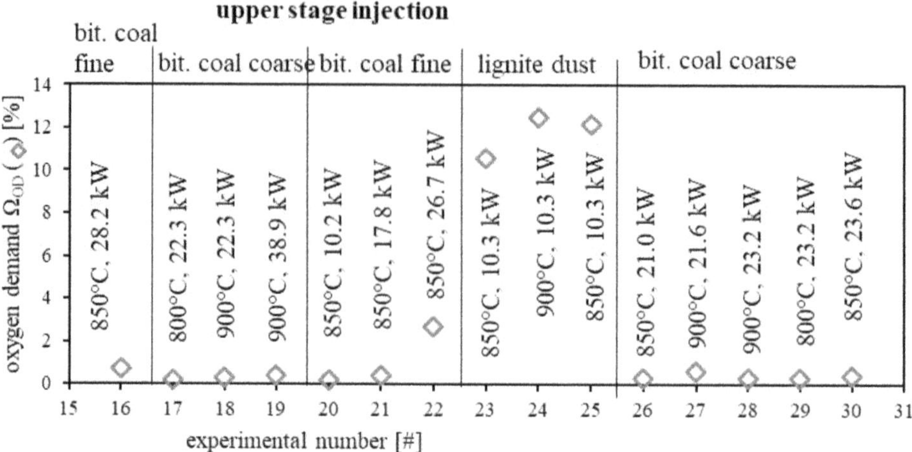


**Figure 43: Oxygen demand experienced during the experimental campaigns, when solid fuel was injected into the upper stage.**

Therefore, coarse lignite was injected again on the lower stage and the results of this campaign is plotted in Figure 44. This again led to low oxygen demand values for this fuel. When methane was inserted alongside the fluidization gases into the reactor, higher concentrations of methane left the reactor unconverted already at a low fuel mass flow of 13.8 kW. This can be traced back to the formation of solid free bubbles at the distributor, which will bypass the reactive bed without converting the methane. In experiment #38 with lignite dust, when the dynamic operation with lignite dust was investigated, similar results than seen before for the oxygen demand could be established, showing that the experiments can be reproduced. Biomass with its high content of volatiles showed relatively high values of oxygen demand for this reactor setting of around 2%.

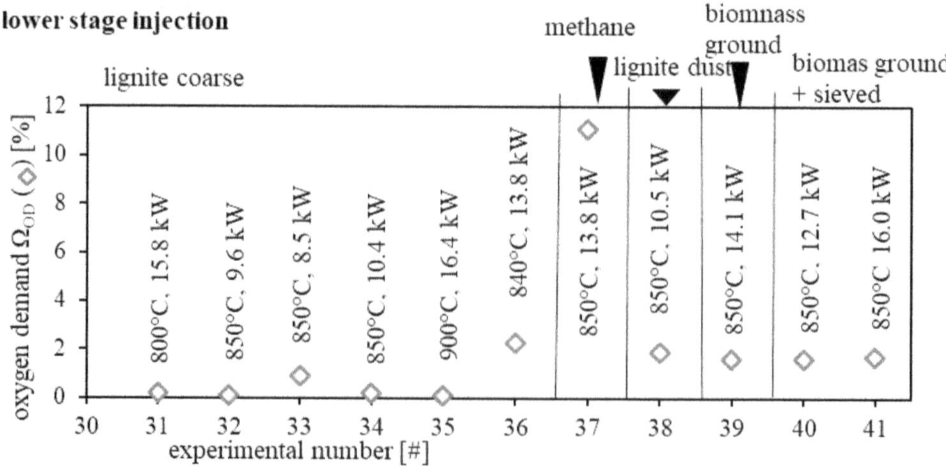


**Figure 44: Oxygen demand experienced during the experimental campaigns, when solid fuel was injected into the lower stage.**

From other researchers one can find that with CLOU, due to the release of gaseous oxygen, the gas conversion is usually close to completion [58,60]. Nevertheless, OCs for CLOU operation are usually highly manufactured and, hence, expensive materials. Under iG-CLC conditions, typical values for the oxygen demand of $\Omega_{OD}$ = 10% - 25% were found by Chalmers University of Technology shown in chapter 2.2.3. Pilot scale operations at CSIC Zaragoza by Perez Vega et al. [57] could be established as low as $\Omega_{OD}$ = 6.5% for bituminous coal. Wang et al. [61] reported around $\Omega_{OD}$ = 7% using bituminous coal. The oxygen demand was significantly higher, when fuels with a higher volatile content were used. This means that the achieved $\Omega_{OD}$ values in the two-stage system, especially for biomass and lignite, are standing out compared to operation of one-stage systems at other institutes.

### 4.1.3. Influence of the second stage on the gas conversion

For the evaluation of the performance parameters of the CLC process, averaged outlet gas concentrations were evaluated. The gases were extracted at the outlets of both AR and FR as described in Figure 14. It is possible to extract gases from the plant already at the end of the freeboard of the first stage in the FR. Comparing the concentrations after the second stage and at the end of the first stage gives a good insight on how the fuel is converted by the two-stage FR design. In the gas concentration section, some of the extractions from the freeboard of the first stage are plotted along the off-gas concentrations. What is clearly visible is that the concentration of combustible gases is always higher above the lower stage than in the outlet of the reactor, which shows that the two-stage design of the FR is sufficing in converting gases. This statement can be emphasized by looking at the previously determined oxygen demands, which for solid fuels were never above 2% in the system.

How the various fuel types and their size fractions profit from the design can be read by the concentrations after the first stage and at the outlet of the reactor summarized in Table 11. Fine lignite dust showed extremely high concentrations of combustible gases of 16.4 vol.% $H_2$, 1.9 vol.% $CH_4$ and 9.4 vol.% CO, especially considering that a considerable amount of $CO_2$ is used for fluidization and injection. This $CO_2$ dilutes the gas concentrations, which is why the molar balancing for the performance parameters gives more comparable results. Nevertheless, the second stage reacts those combustible gases to very low outlet concentrations. Coarse lignite did not show such high concentrations after the first stage of the FR, which can be either due to the low fuel input at this experiment or the fact that the coarser particles might be located lower inside the bed.

Fine bituminous coal showed some improvements in gas conversion, where $H_2$, CO and $CH_4$ went down by some percent at 850°C and 900°C. This fuel showed relatively low concentrations of those gases after the first stage, which brought up the idea to inject this fuel into the upper stage. Methane profited the most from the two-stage design, where the concentration above the first stage was still at 35.7 vol.% from around 62 vol.% at the inlet. The 2$^{nd}$ stage converted methane to the exit concentration of 4.7 vol.%. Biomass showed in general strong improvements of gas conversions over the 2$^{nd}$ stage by several percent for each of the combustible gases. This effect is not prominently visible in Table 11, because of the low fuel mass flows and the high amount of $CO_2$ stemming from injection and fluidization gas. Another point to mention is the increasing leakage flow of oxygen/air in course of the experiments. This indicates the need for thorough check-up of the whole plant to see, whether gas sealing parts inside the system or towards the gas concentration measurements are working properly.

**Table 11: Comparison of 1st and 2nd stage extraction of FR gases.**

| Fuel | gas measurement at the exit of | fuel power $kW_{th}$ | $H_2$ vol.% | $CO_2$ vol.% | $CH_4$ vol.% | CO vol.% | $O_2$ vol.% |
|---|---|---|---|---|---|---|---|
| lignite fine 850°C | 1st stage | 18.7 | 16.4 | 72.3 | 1.9 | 9.4 | 0 |
| exp. #13 | 2nd stage | | 1.4 | 97.2 | 0 | 0.8 | 0.5 |
| lignite coarse 850°C | 1st stage | 8.5 | 0.9 | 84.7 | 0 | 0.7 | 2.1 |
| exp. #33 | 2nd stage | | 0.0 | 82.3 | 0 | 0.1 | 3.0 |
| bit. fine 850°C | 1st stage | 17.6 | 1.6 | 97.0 | 0.3 | 1.1 | 0 |
| exp. #14 | 2nd stage | | 0.7 | 98.8 | 0 | 0.1 | 0.4 |
| bit. fine 900°C | 1st stage | 20.7 | 2.8 | 94.7 | 0.3 | 2.2 | 0 |
| exp. #15 | 2nd stage | | 0.7 | 98.9 | 0 | 0.1 | 0.3 |
| Methane 850°C | 1st stage | 15.3 | 1.1 | 61.3 | 35.7 | 0.4 | 2.1 |
| exp. #38 | 2nd stage | | 0.7 | 86.4 | 4.7 | 0.4 | 3.2 |
| biomass coarse 850°C | 1st stage | 14.0 | 2.1 | 98.8 | 0.9 | 1.9 | 4.2 |
| exp. #39 | 2nd stage | | 0.5 | 78.4 | 0 | 0.1 | 5.3 |
| biomass fine 850°C | 1st stage | 12.7 | 1.8 | 84.1 | 0.3 | 1.2 | 3.0 |
| exp. #40 | 2nd stage | | 1.1 | 76.6 | 0 | 0.1 | 5.1 |
| biomass fine 850°C | 1st stage | 16.0 | 4.0 | 77.8 | 1.8 | 5.8 | 2.4 |
| exp. #41 | 2nd stage | | 1.3 | 80.1 | 0 | 0.2 | 4.1 |

### 4.1.4. Dynamic operation of the pilot plant

Methane feeding

Load changes, as well as start and stop of fuel injection were carried out during the operation with methane, lignite dust, and bituminous coal as fuels. The gas concentration measurements for the fuel starts and stops conducted with methane are shown in Figure 45. Methane is shown first, because it shows no solid holdup being gaseous. From these measurement results it was estimated, how long the total change from one steady state result to another takes timewise. The FR exitconcentrations show that after several minutes the methane front with the final outlet concentration is visible. When fuel is not injected the $CO_2$ concentration slumps, because of the higher proportion of leakage gases. Generally, the dynamics are better investigated by looking at the AR off-gas concentrations. This was done by checking when the $O_2$ concentration in the AR reaches the first time 20.9 vol.% after a fuel stop and 14.5 vol.% after the fuel start, which is a long-time average value for the used fuel input. Time dependent behavior of the measurement devices was disregarded, because it is assumed that both start and stop will show the same delay in concentrations. For the gaseous fuel methane, it was extracted that the fuel stop induced a 10 minutes total reaction of the system, whereas with starting the injection the system needed 12 minutes long.

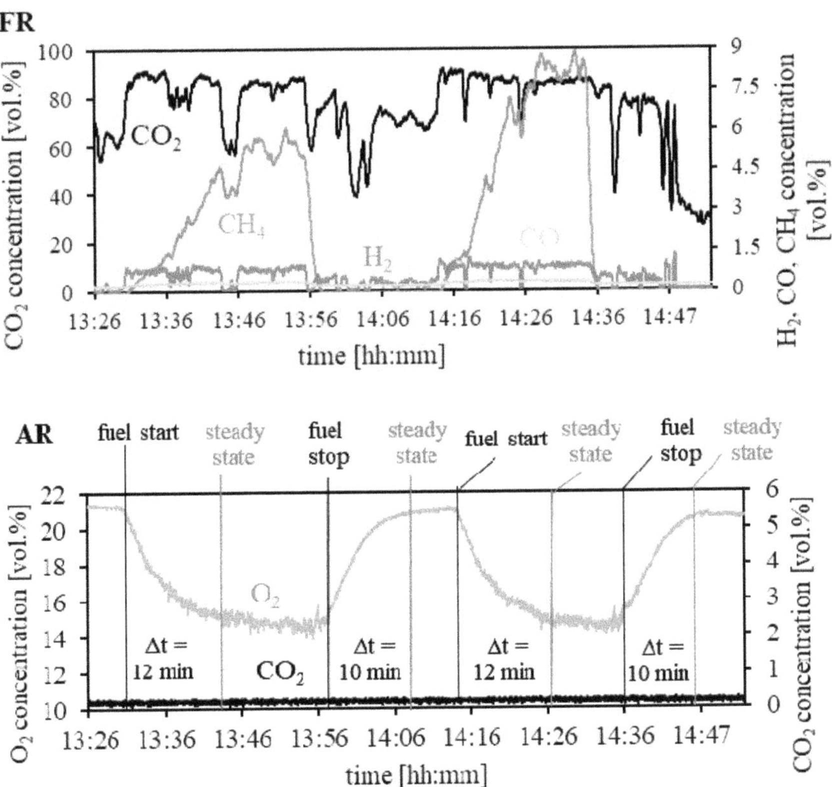


**Figure 45: Dynamic changes in dry gas concentration in the off-gas due to fuel start and fuel stops conducted with methane (15.3 kW$_{th}$, 850°C).**

Fine Bituminous Coal

To increase char conversion, fine bituminous coal was injected into the upper stage of the fuel reactor system in experiments #20-#22. Within these experiments, fuel starts and stops as well as load changes could be accurately operated and are plotted in Figure 46. Fuel start and fuel stop with 10.2 kW fuel power yielded response times of the system of around 10 min. The load change from 10.2 kW to 17.8 kW took longer with around 13 min. When the fuel injection was increased to 26.7 kW, the change took around 18 min until the changes in AR gas concentrations stabilized. The following fuel stop took 10 min, like the fuel stop from 10.2 kW. It can be clearly stated that the load changes had a longer response time, the more fuel was injected beforehand. This might be traced back to the carbon hold-up in the FR, which seems to find a new equilibrium in a bigger timeframe, the more carbon is inserted.

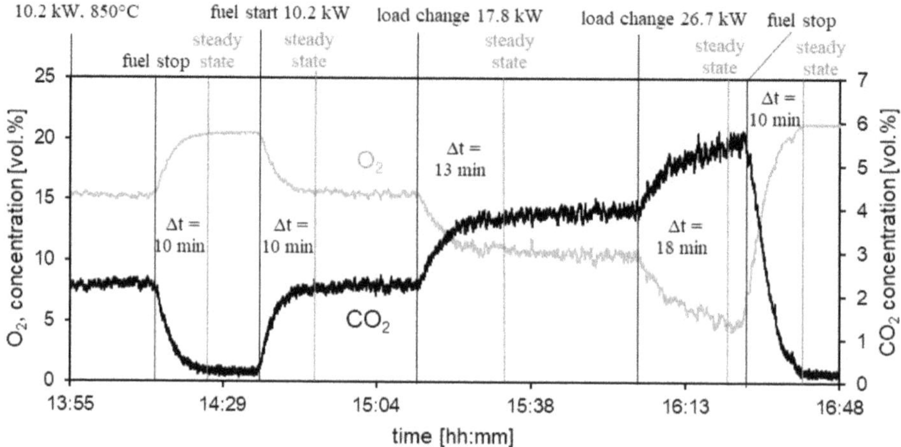


**Figure 46: Dynamic changes in dry gas concentration in the off-gas of the AR due to fuel start, load changes and fuel stops. Experiments conducted with fine bituminous coal injected into the upper stage of the FR system (850°C).**

Ground and sieved Biomass

The next fuel, which is shown in this segment, is ground and sieved biomass. The biomass was assumed to have a low carbon hold-up during conversion due to the high volatile content. In Figure 47, the transient change in $CO_2$ and $O_2$ gas concentration are plotted, alongside the applied changes to the system. A major difference compared to methane is the stronger fluctuations in the gas concentrations, which can be traced back to the unsteady fuel supply with the fuel feeder. Nevertheless, by averaging the $O_2$ over a time of 20s, the same criteria for determining the timeframes of the changes in the system can be applied. Interestingly, the time scales of the transient changes are interchanged, the fuel stop takes around 12 minutes, whereas the fuel start is taking around 10 minutes in the system. One explanation can be the accumulation of some carbon hold-up during fuel injection, which can be read from the elevated $CO_2$ concentration in the AR. This will lead to a prolonged period, in which this carbon is moved and converted out of the FR lower bed. Other effects, which have a major impact, like solid circulation, fuel gases and their effect on the hydrodynamic need to be evaluated by detailed modeling as they cannot be extracted directly from the gas concentration measurements.

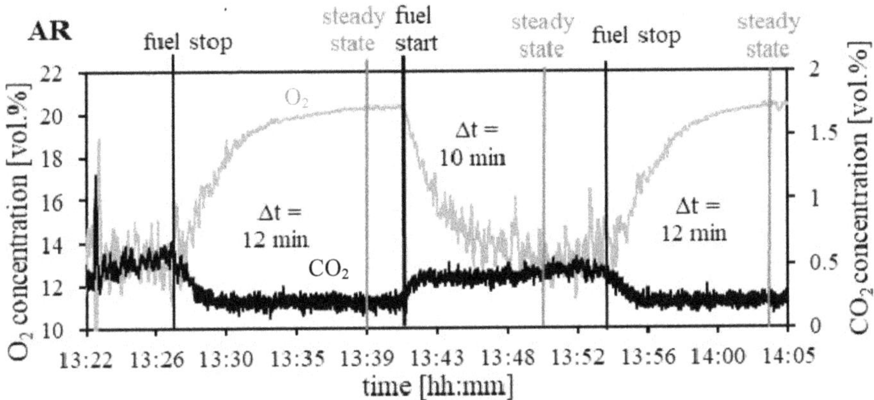


**Figure 47: Dynamic changes in dry gas concentration in the off-gas due to fuel start and fuel stops conducted with ground and sieved biomass (16.0 $kW_{th}$, 850°C).**

## 4.2. Experimental lab scale fluidized bed

The experimental lab scale facility was used to investigate the complex conversion mechanism of lignite char in a bed of OC particles. First the gas concentration measurements are analyzed, and later detailed reaction kinetic parameter extraction is conducted.

### 4.2.1. Gas concentration measurements

Gas concentrations were collected from all experiments. By looking at this data alone, it is already possible to see changes in the reaction mechanisms among different operation conditions.

Inert bed gasification

Steam and $CO_2$ gasification were examined in an inert bed first, to get a reference reaction behavior and a possibility for comparison with oxygen carrier experiments. Figure 48 shows exemplarily the gasification behavior of the size fraction with $d_S = 483$ µm in an experiment with $CO_2$ only as gasification agent (a) and with steam injection (b) only. The gasification with pure carbon dioxide follows solely equation (1) as no other reactants are present. When only steam is used, the gasification is taking place via equation (2), but also the WGS reaction (3) is obviously taking place, because significant carbon dioxide concentrations are visible. The carbon dioxide itself works as a gasification agent again following equation (1), which in the end explains the high CO concentrations during steam gasification. With a total conversion time of 500 s, the steam gasification is faster at this setting, because for pure $CO_2$ the gasification takes 800 s. This is similar to results found by other researchers for the inert bed gasification of lignite [134].

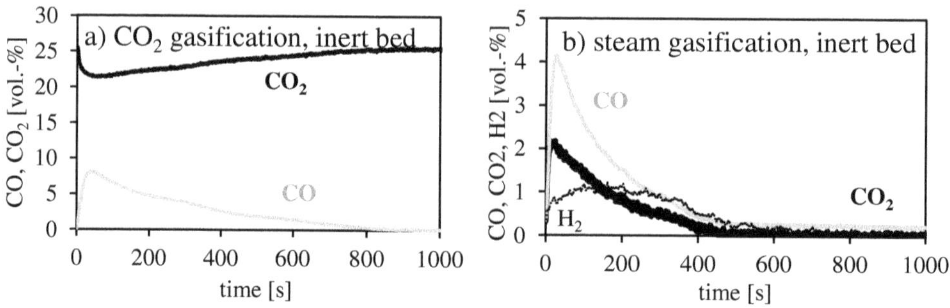


**Figure 48: Gas concentration measurements (dry) during the gasification of 1g lignite char, $d_S = 483$ µm, at 1000 °C, $u_R = 0.165$ m/s. Gasification agent is CO2 (a) and steam (b) with an inlet concentration of both at 25 vol.-%.**

Reactive bed gasification

Exchanging the inert bed with a reactive OC bed leads to remarkable changes in the conversion behavior. In Figure 49 (a) it is visible that the previously observed consumption of CO2 shifts to generation of the very same gas, when lignite is gasified with CO2. The oxidation of the gasification product CO by the oxygen carrier on the one hand and the generation of the reactant CO2 on the other hand, leads to a significant decrease of the time for total gasification. Steam gasification in Figure 49 (b) of the coarse lignite fraction of 1409 µm in a reactive OC bed is also remarkably faster than the conversion in an inert bed. This is visible from the considerably higher concentrations of outlet carbonaceous gas streams in Figure 49 (b) compared to the previous inert bed experiments.

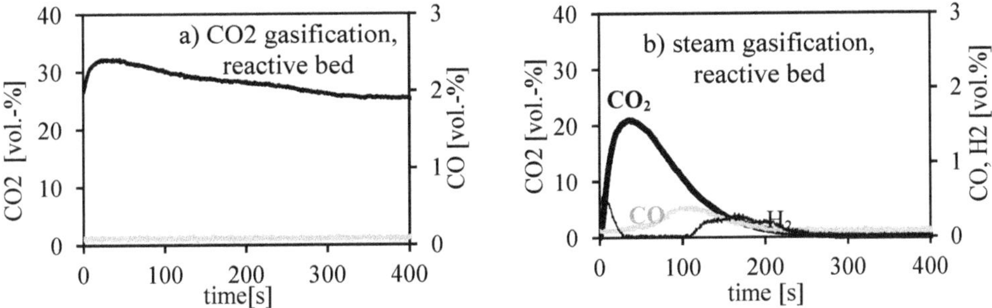


**Figure 49: Gas concentration measurements (dry) during the reactive bed gasification of 1g lignite char, a) $d_S$ = 1251 µm b) $d_S$ = 1409 µm, at 900 °C, $u_R$ = 0.15 m/s. Gasification agent is CO2 (a) and steam (b) with an inlet concentration of both at 25 vol.-%.**

Inhibition of inert sand bed gasification by hydrogen and carbon monoxide

The inhibition of the gasification can be made visible only in an inert bed, because the oxygen carrier would convert the inhibition gases CO and $H_2$ steadily. In Figure 50, the gas concentrations during the experiments without and with the addition of 15 vol.-% CO and 10 vol.-% $H_2$ are plotted. The addition of CO to the gasification agent will make the reaction about half as fast, which can be read from the lower CO that is generated because of the reaction itself. When hydrogen is added, the reaction pattern becomes more complex. Introducing 10 vol.-% hydrogen will lead to a maximum of 1 vol.-% measured at the outlet, meaning that the remainder is reacting with the $CO_2$ to CO and water via WGS. From a molar balance of all carbonaceous flows, one can find that the gasification is finished at around 1200 s compared to 600 s without addition of hydrogen.

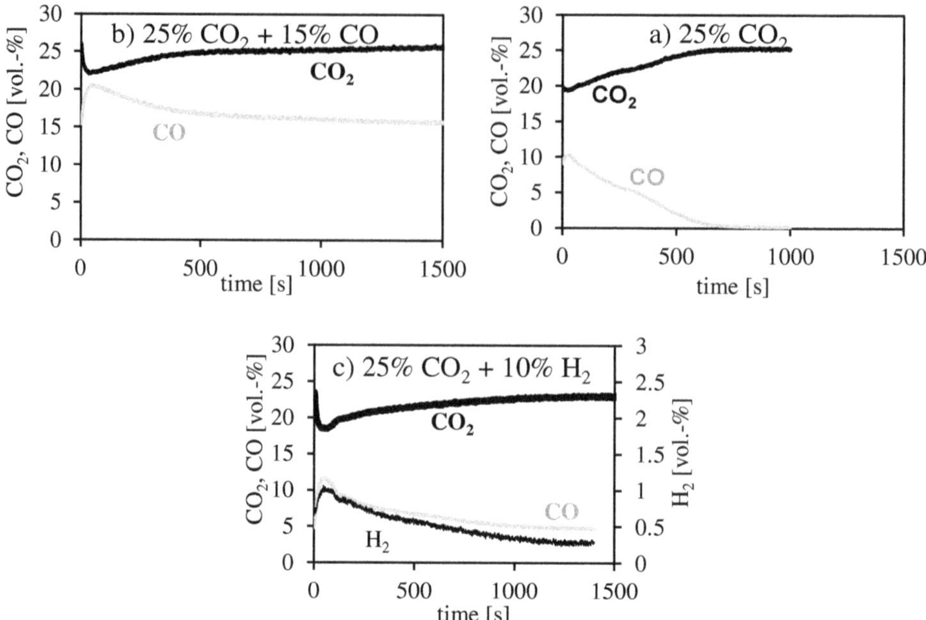


**Figure 50: Gas concentration measurements (dry) during the gasification of 1g lignite char, $d_S = 871$ µm, at 900 °C, $u_R = 0.15$ m/s. (a) Gasification agent is $CO_2$ with an inlet concentration of 25 vol.-%. In separate tests, (b) 10 vol.-% $H_2$ and (c) 15 vol.-% CO were added to the fluidization gas.**

Influence of the fuel particle size on reactive bed gasification

In inert bed experiments it could be shown that the particle size of the coal char was not a major parameter for the conversion kinetics. Only above a mean particle size of 1 mm, a longer conversion time, meaning a longer duration of elevated $CO_2$ levels, was observed. This behavior changes dramatically for the reactive bed experiments, which are depicted in Figure 51. The total time of conversion is around 150 s for the smallest particle size investigated of $d_S = 173$ µm and rises to around 350 s for the largest investigated particles with $d_S = 1251$ µm. For smaller particle sizes below 871 µm, CO could be detected during gasification, especially high levels for the smallest investigated fraction. A possible explanation for the presence of CO can be found in the fluidization behavior of smaller particles, which tend to float more in the upper part or on top of the fluidized bed. There, the produced carbon monoxide does not come into contact with oxygen carrier enough to be fully converted.

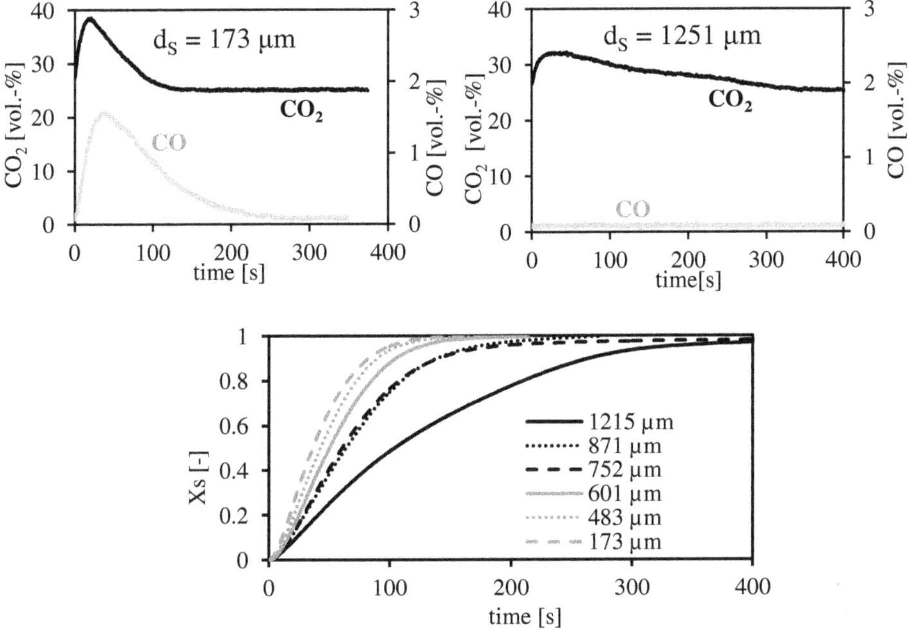


**Figure 51: Top: Gas concentrations (dry) during the gasification of 1g lignite char for two different size fractions; below: solid conversion Xs over time for all investigated particle size fractions, at 900 °C, $u_R$ = 0.15 m/s. Gasification agent is CO2 with an inlet concentration of 25 vol.-%.**

Simultaneous $CO_2$ and steam gasification

In a CLC system it is to be expected that the OC bed in the fuel reactor will be fluidized with recirculated gases consisting of steam and carbon dioxide. This means that the gasification of the solid fuel will happen in an atmosphere of both gases. Experiments with different concentrations of $CO_2$ and $H_2O$ were conducted to check, which reaction path is predominant or if they influence each other. Data for inert and reactive bed simultaneous gasification experiments are plotted in Figure 52. In the inert bed case, carbon dioxide is generated, meaning that the steam reacts with the CO from the gasification to form hydrogen and $CO_2$ via the WGS reaction. It cannot be retrieved directly from the data to which degree steam and $CO_2$ gasification are responsible for the reaction with the carbon directly. This is due to the fact that both reactions are interconnected via the WGS reaction on both the reactant and the product side. The same is true for the reactive bed experiments, but one can see that the gasification reaction is enhanced by the presence of the OC. Further, the combustible gases CO and $H_2$ are converted to a high degree to water and $CO_2$, which can be seen at the high $CO_2$ peak.

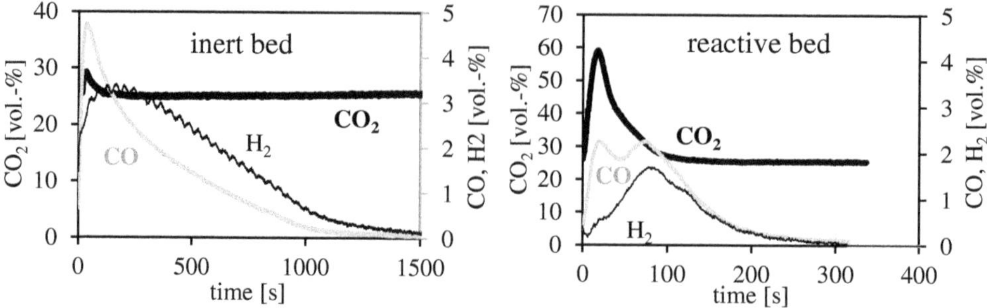


**Figure 52: Gas concentration (dry) during the gasification of 1g lignite char, at 900 °C, $d_S$ = 334 μm, $u_R$ = 0.15 m/s in an inert bed (left) and a reactive bed (right). Gasification agent is a mix of 25 vol.-% steam and 25 vol.-% $CO_2$. The time axis was varied to give a better resolution for both experiments.**

### 4.2.2. Conversion behavior analysis

The data evaluation showed that the SCM generally achieved better results than the VRM. Therefore, only the kinetic parameters calculated for the SCM, $k_{SCM}$, are plotted in the following.

<u>Gasification with either $CO_2$ or steam</u>

As a reference case, lignite char was gasified in an inert sand bed. In Figure 53, the influence of particle size, temperature and gas concentration on the steam gasification as well as on the $CO_2$ gasification is shown. Higher temperatures enhance the gasification considerably throughout all experiments with steam and also with $CO_2$. Steam gasification is slightly faster than with $CO_2$, especially at 1000°C. Looking at the $CO_2$ gasification in detail reveals that the particle size does not have measurable effects on the gasification rate for particle Sauter diameters below 850 μm. When using steam as reacting agent, it turns out that smaller particles increase the gasification rate throughout the experiments. This leads to the conclusion that the $CO_2$ gasification rate is mostly controlled by the chemical reaction, whereas the steam gasification is both influenced by diffusion as well as by chemical reactions inside the char. For the steam gasification, the resistance approach from equation (59) captured the particle size influence on the gasification kinetics well.

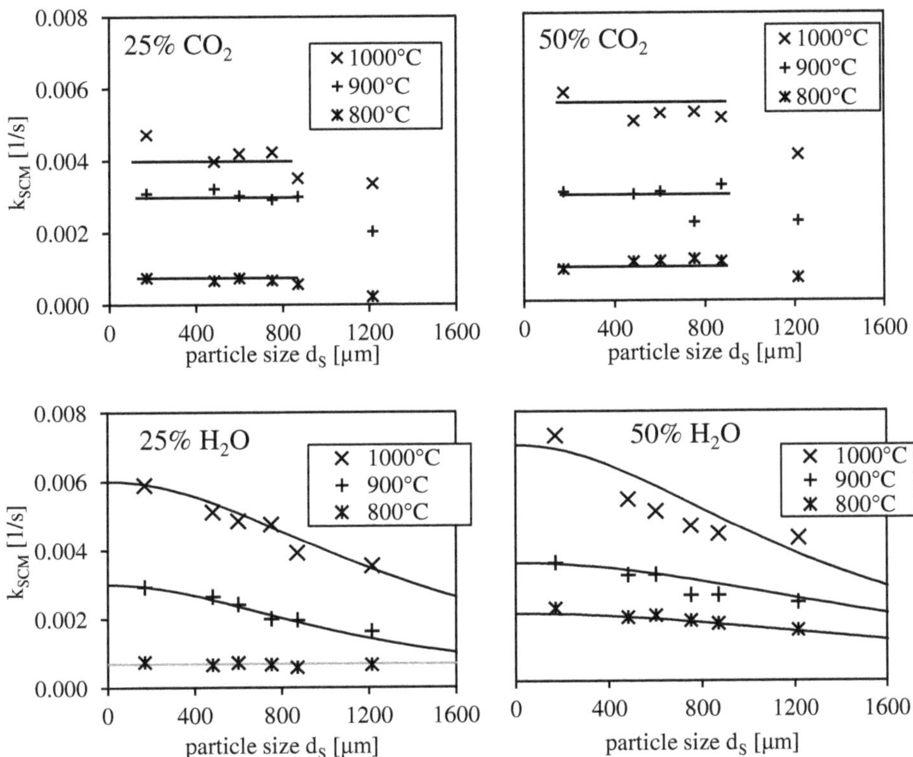


**Figure 53: Char gasification with $CO_2$ or steam in the inert bed. (Kinetic constants derived from particle conversion analysis with SCM for steam and $CO_2$ gasification. Gasification experiments in an inert sand bed. The lines show kinetics calculated with the resistance approach from equation (59), which is described in detail in section 3.2.3)**

Exchanging the inert sand bed with a bed of OC material, the gasification reaction is accelerated drastically. This is demonstrated in Figure 54, where the pure steam and $CO_2$ gasification reactions are compared for the setups with and without OC. For coarser particles above 1 mm, the OC accelerates the gasification rate three times for pure $CO_2$ and around four times for pure steam. Smaller char particles profit even more from the presence of OC and the gasification is six times enhanced for $CO_2$ and tenfold for steam.

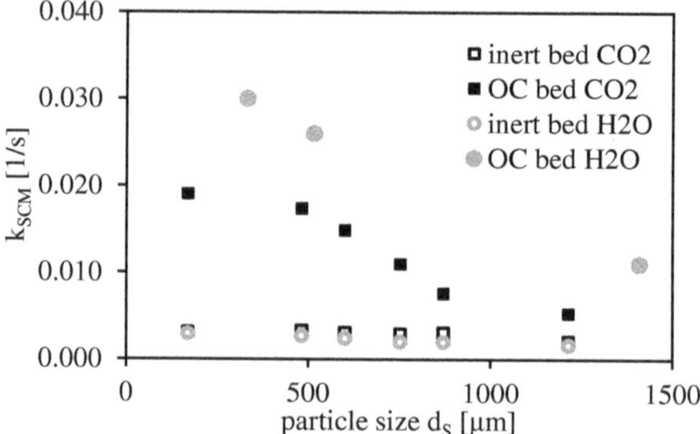


**Figure 54: Char gasification with $CO_2$ or steam in the inert bed and in the OC bed, respectively. (Kinetic constants derived from particle conversion analysis with SCM for 25 vol.-% steam and 25 vol.-% $CO_2$ gasification, respectively. Gasification experiments in an inert sand bed compared to OC bed results. T = 900 °C, $u_R$ = 0.15 m/s.)**

Inhibition of gasification by carbon monoxide and hydrogen addition

As it has been shown above, the OC increases the gasification rate. An explanation could be that the OC oxidizes the gasification products CO and $H_2$ to $CO_2$ and steam and, with it, influences the chemical equilibrium. The steam and $CO_2$ can then react as gasification agents themselves and further increase the gasification rate. If this mechanism is valid, the addition of $H_2$ and CO to the gasification agent should lead to a decrease in gasification activity in the fluidized bed. This is confirmed by the data shown in Figure 55, where it becomes clear that the presence of both inhibiting gases decreases the gasification rate by 50-60%.

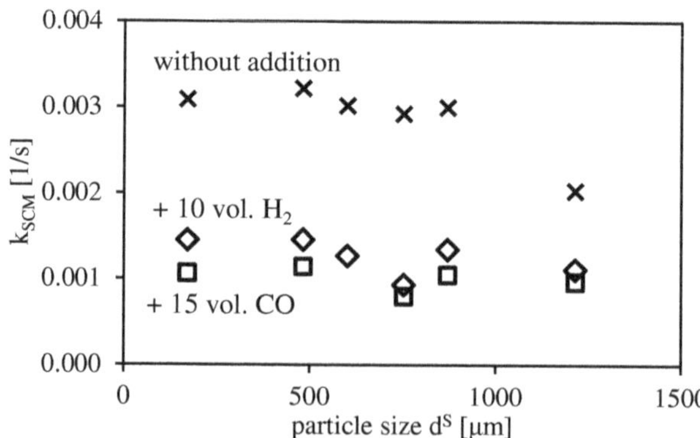


**Figure 55: Inhibition of char gasification by addition of CO and $H_2$, respectively. (Kinetic constants derived from particle conversion analysis with SCM for 25 vol.-% $CO_2$ gasification experiments. Gasification experiments in an inert sand bed with and without addition of $H_2$ and CO as inhibiting gases. T = 900 °C, $u_R$ = 0.15 m/s.)**

Gasification under iG-CLC conditions: simultaneous $CO_2$ and steam gasification

Figure 56 shows that gasification with a mixture of $CO_2$ and steam is faster than $CO_2$-only gasification but does not enhance gasification compared to pure steam gasification. In the Figure one can also find the simultaneous gasification rates in an inert bed, which are much lower than the reactive bed ones. In the inert bed experiments the same effect about the gas mixtures was visible; the simultaneous gasification with 25 vol.-% steam and 25 vol.-% $CO_2$ does not exceed the one with steam only. This means that the presence of steam plays the major role in iG-CLC conditions and for the used fuel. In addition to that, it becomes clear that in iG-CLC with lignite and a CuO-based oxygen carrier, the fuel particle size is important for the conversion of the fuel, because throughout the experiments in a reactive bed, the rates for finer particles were above the ones of the coarser size fraction.

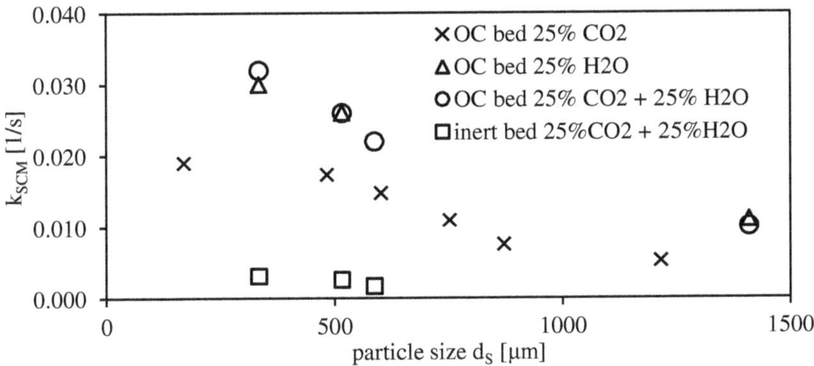


**Figure 56: Char gasification with gases present during iG- CLC operation. (Kinetic constants derived from particle conversion analysis with SCM for experiments with $CO_2$ and steam in OC and inert bed, respectively. T = 900 °C, $u_R$ = 0.15 m/s.)**

Gasification under iG-CLC conditions: variation of $CO_2$ and steam concentrations

The calculation of the kinetic constant $k_{SCM}$ for simultaneous injection of $CO_2$ and steam showed that the gasification rate equals roughly the steam-only rate. This could mean that an increased concentration of steam could further improve the gasification in the fluidized bed reactor. This was tested in a series of experiments with different inlet concentrations of $CO_2$ and steam. The results are depicted in Figure 57 and show that the highest steam concentration resulted in a slightly elevated gasification rate compared to lower concentrations. The smallest particle size profits most from the increased steam concentration, which indicates a diffusion limitation for the whole investigated size range. For the biggest size fraction, the acceleration due to higher steam concentrations was not observed.

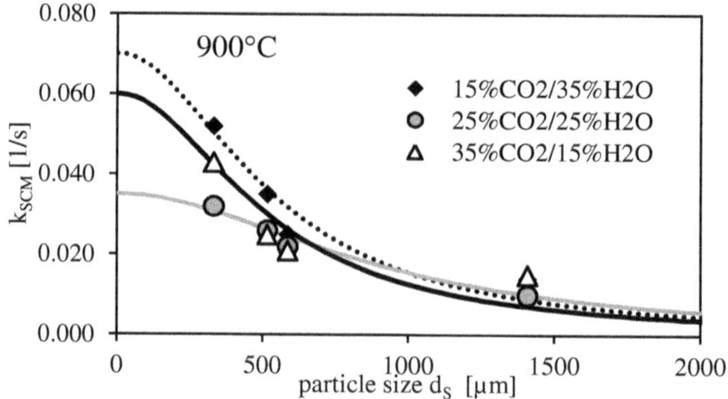


**Figure 57: Influence of varied gas concentrations during char gasification. (Kinetic constants derived from particle conversion analysis with SCM for simultaneous $CO_2$ and steam gasification experiments in an OC bed. Gas concentrations varied. T = 900 °C, $u_R$ = 0.15 m/s. The lines show calculated kinetics with the resistance approach from eq. (20), which is described in detail in section 4.3.)**

<u>Gasification under iG-CLC conditions: variation of operating velocity $u_R$</u>

Another important parameter, which could influence the conversion of char during gasification, is the superficial velocity $u_R$ in the reactor. It influences mass transport around the particles and, additionally, the velocity changes the fluid dynamics in a fluidized bed reactor with respect to bubble formation. In Figure 58, a series of experiments dedicated to the fluidization velocity during simultaneous $CO_2$ and steam gasification is shown. From the results, it can be concluded that the fluidization velocity does not influence the reactions significantly.

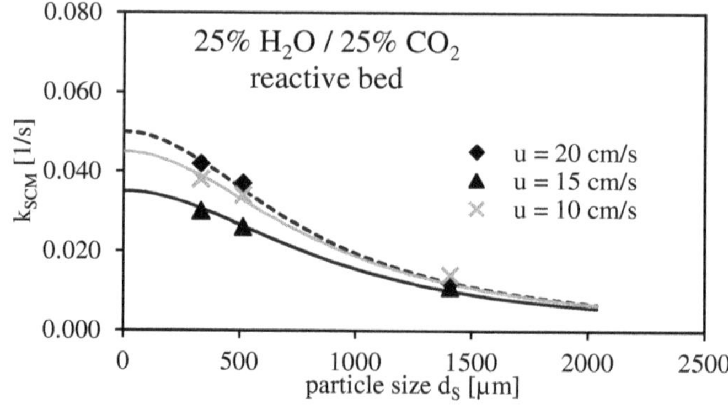


**Figure 58: Influence of the reactor velocity $u_R$ on char gasification in the OC bed. (Kinetic constants derived from particle conversion analysis with SCM for simultaneous $CO_2$ and steam gasification experiments in an OC bed. T = 900 °C. The lines show calculated kinetics with the resistance approach from eq. (20), which is described in detail in section 4.3.)**

<u>Gasification under iG-CLC conditions: temperature dependence of simultaneous $CO_2$ and steam gasification</u>

Finally, the influence of the temperature on the char conversion was scrutinized. In Figure 59, it is shown how much the gasification rate profits from elevated temperatures. Again, finer particles are converted faster than coarser particles in otherwise identical operation conditions. For the two temperatures 900 °C and 1000 °C the reaction shows a diffusion limitation over the whole investigated size spectrum of char particles.

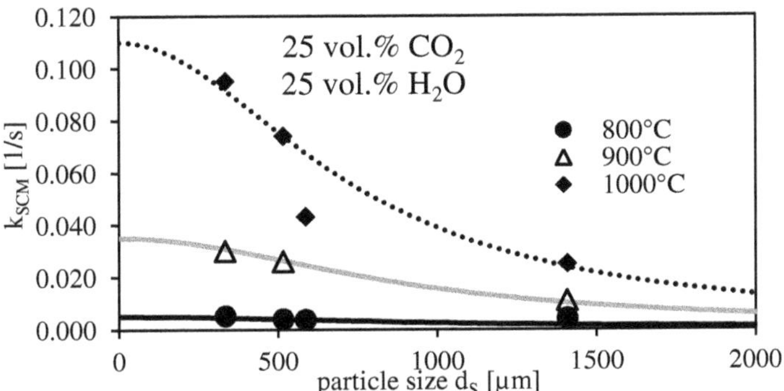


**Figure 59: Temperature variation during char gasification in an OC bed. (Kinetic constants derived from particle reaction modeling via SCM for simultaneous $CO_2$ and steam gasification experiments in an OC bed. The lines show calculated kinetics with the resistance approach from eq. (20), which is described in detail in section 4.3.)**

### 4.2.3. Reaction kinetics parameters

The conversion analysis revealed that several factors influence the conversion of char particles in a CLC environment. The presence of a reactive bed of OC material accelerated the gasification compared to the inert bed $CO_2$ and $H_2O$ conversion tremendously, but one major impact can also be attributed to the particle size. With the help of the resistance approach defined in equation (59), the temperature dependent effect of the chemical reactions as well as the particle size dependent diffusion can be separated. Figure 20 shows the plots to extract the $E_a/R$ of the Arrhenius reaction kinetics. $E_a/R$ is in the range of 11400 – 12000 K for the inert bed gasification, which results in activation energies ranging from 75 – 115 kJ/mol. The reactive bed simultaneous gasification results in an $E_a/R$ of 21200 K, being equivalent to an activation energy of 176 kJ/mol. As is shown in Figure 60, the reaction order for steam gasification in an inert bed is around 0.42. Using $CO_2$ in an inert bed, the reaction order is 0.30. Due to the significant production of $CO_2$ and $H_2O$ during the reactive OC bed experiments, a dependency of the gasification rate on the inlet gas concentrations could not be determined and, hence, the reaction order for reactive bed experiments is assumed to be 0 in the reactive bed.

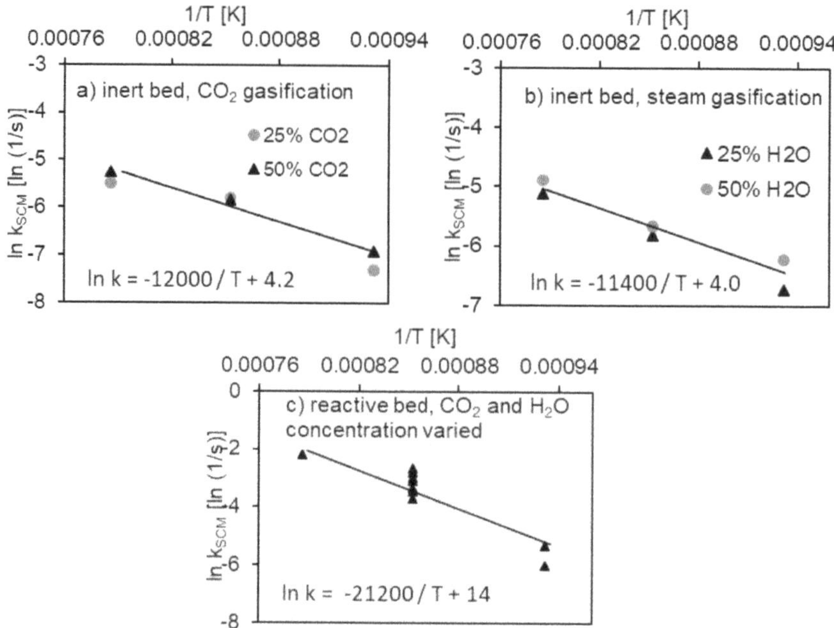


**Figure 60: Arrhenius plots to extract Ea/R of CO2 and steam gasification in both the inert bed and the bed of reactive OC particles, various concentrations of CO2 and/or steam in the reactive OC bed.**

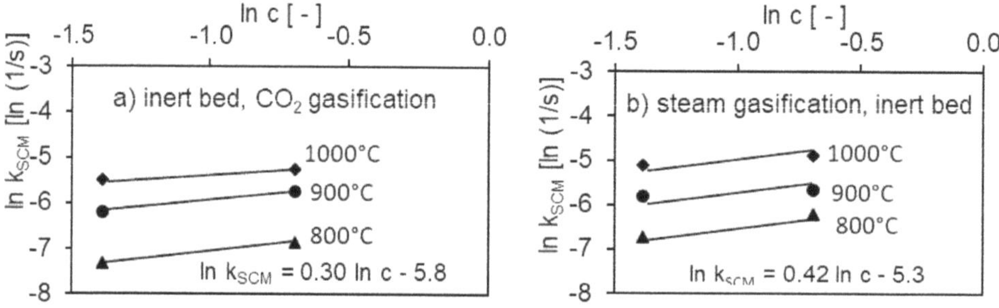


**Figure 61: Kinetics plots for determining the reaction order of steam and CO2 gasification in an inert sand bed.**

The ultimate goal of the present investigation is the extraction of a set of kinetic parameters, which can be used to simulate the gasification of the lignite char for different CLC operation conditions. The kinetic parameters in

Table 12 are proposed for the process modeling of lignite char in both an inert and a reactive fluidized bed reactor.

**Table 12: Kinetic parameters for the gasification reaction during $CO_2$ and steam gasification using an inert or a reactive CuO-based OC fluidized bed in the chemically controlled regime.**

| | agent type | $E_a$ [kJ/mol] | $k_0$ [$s^{-1}$] | n [-] |
|---|---|---|---|---|
| inert bed | $CO_2$ | 100 | 94 | 0.30 |
| | $H_2O$ | 95 | 85 | 0.42 |
| reactive bed | $CO_2$ | 176 | $1.7 \cdot 10^6$ | 0 |
| | $H_2O$ | 176 | $2.2 \cdot 10^6$ | 0 |
| | $CO_2$ + $H_2O$ | 176 | $2.2 \cdot 10^6$ | 0 |

These parameters can be used if particle sizes are gasified, where no diffusion limitation occurs, and the rate is governed by chemical reaction. For bigger particles an effective diffusivity $D_e$ according to equations (59) and (60) has to be applied to describe the diffusion effects on the gasification rate in a reactive oxygen carrier bed. The effective diffusivity of the simultaneous $CO_2$ and steam gasification in a reactive fluidized bed is dependent on the temperature, which can be seen in results of Figure 62. From these experiments the linear relationship in Figure 63 could be extracted, which describes the diffusion influence in the examined temperature span from 1073 K to 1273 K.

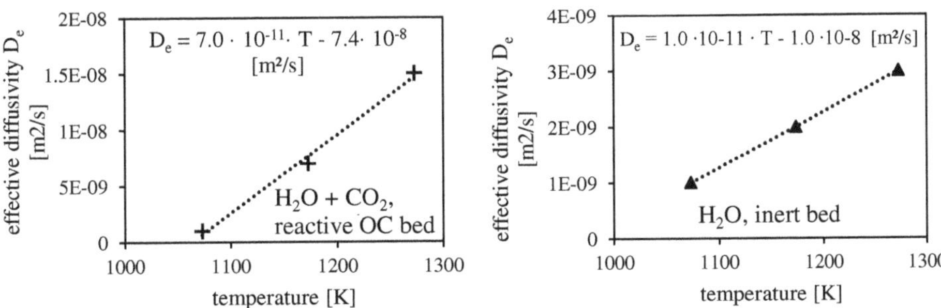


**Figure 62: The effective diffusivity in the temperature range of 1073 to 1273 K for steam gasification in an inert fluidized bed and for the simultaneous CO2 and steam gasification in a reactive OC fluidized bed.**

In a final step, these parameters are used to calculate the conversion of the char particles. In Figure 63 it is demonstrated how this approach is able to describe the experimentally seen conversion accurately in a broad spectrum of sizes and temperatures under iG-CLC conditions.

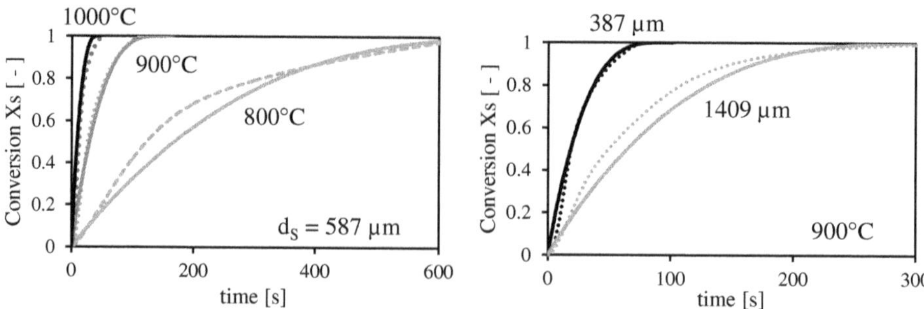


**Figure 63: The effective diffusivity in the temperature range of 1073 to 1273 K for steam gasification in an inert fluidized bed and for the simultaneous $CO_2$ and steam gasification in a reactive OC fluidized bed. Dotted plots show the experimental data, continuous plots show the corresponding simulation results.**

## 4.3. Dynamic Flowsheet Simulation results

Gas composition and situation in the FR stages

In the experimental campaign with methane, the volumetric gas concentration after FR stages 1 and 2 were measured. There it was found that the methane conversion was not complete after the second stage. Time-averaged values of several minutes for the concentration measurements are provided in Table 13.

**Table 13: Measured average volumetric gas concentrations during operation with 15 $kW_{th}$ methane. Additionally, the inlet concentration is given. Missing percentages are due to leakage flows of $N_2$ and $O_2$ from the AR.**

| dry gas concentration | unit | $CH_{4exp}$ | $CO_{2exp}$ | $H_{2exp}$ | $CO_{exp}$ |
|---|---|---|---|---|---|
| inlet FR1 | [vol.%] | 60 | 40 | 0 | 0 |
| after lower stage FR1 | [vol.%] | 32 | 64 | 1.1 | 0.5 |
| after upper stage FR2 | [vol.%] | 4.7 | 88 | 0.8 | 0.4 |

Two transient flowsheet simulation runs were carried out for the methane conversion. In the first simulation run, literature kinetics from Abad et al. [78], obtained from TGA measurements, were used. As shown in Figure 64, the lower stage of the FR converts the methane to around 7 vol.% but does not suffice to convert much of the hydrogen and carbon monoxide from the decomposition. Entering the second stage of the FR, the gases are well mixed into a bed of fresh, i.e. highly oxidized OC. In this second stage CO and $H_2$ were converted further to exit levels of 35 vol.-% CO and 20 vol.-% $H_2$. Compared to the simulations of Abad et al., the conversion of the methane is extraordinarily fast at the border between dense bed and freeboard, because from a calculation perspective, the gases and solids are perfectly mixed when they enter the freeboard [78]. Looking at the gas measurements during the experiments, one can see that the methane conversion in the simulation is overestimated in both stages. For carbon monoxide and hydrogen, it is the other way around and the conversion found during the experiment of the two gases is much larger than the simulations show.

The region directly above the dense bottom bed with a relatively high solids concentration leads to a jump in overall gas conversion. The sharp change from dense bed to freeboard will in practice be replaced by a transition zone, i.e. a splash zone [135], where the bubbles burst. In the dense bed itself, the methane conversion is comparatively low due to mass transfer limitations by the formation of bubbles.

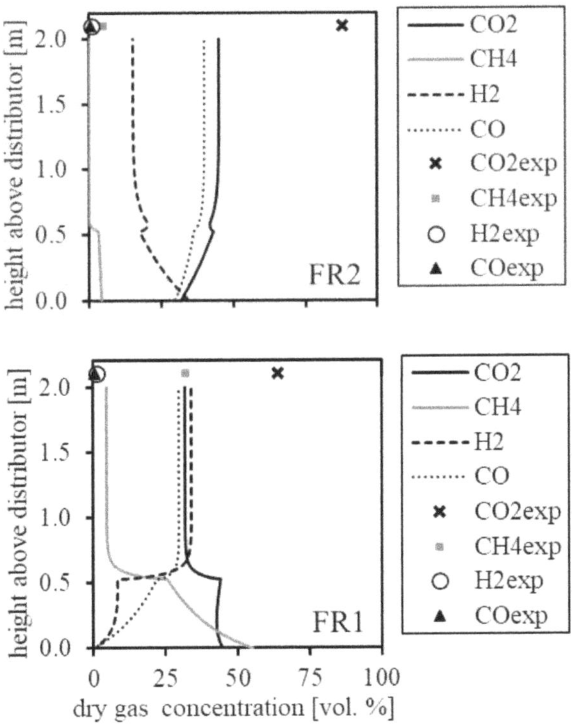


**Figure 64: Simulation of methane conversion inside the FR system with literature kinetics. Shown concentrations in the lower dense zone are averaged over the bubble and suspension phase. Comparison to averaged gas concentration measurements at the exit of each stage.**

In a next step, the kinetics for the methane conversion by CO and $H_2$ from literature were varied to match the experimental results better. This was done by multiplying the kinetic parameter k in equation (15) with a certain factor for each gas species separately. This simple fitting of the kinetic constant k demonstrates in Figure 66 that the exit concentrations can be met. The kinetic parameter for methane conversion is cut in half and the parameters for the $H_2$ and CO are multiplied by the factor 3 and 5, respectively. The need for altering the kinetic parameters shows that the TGA kinetics of a small number of particles cannot be directly transferred to the modeling of a pilot scale system with tens of kilograms of bed material. The decrease of the kinetic parameter for methane shows that there is a mass transfer limitation towards the particles in the fluidized bed. The enhanced rates for carbon monoxide and hydrogen, which originate from the methane, show that they are directly converted further to $CO_2$ and $H_2O$ inside the particle.

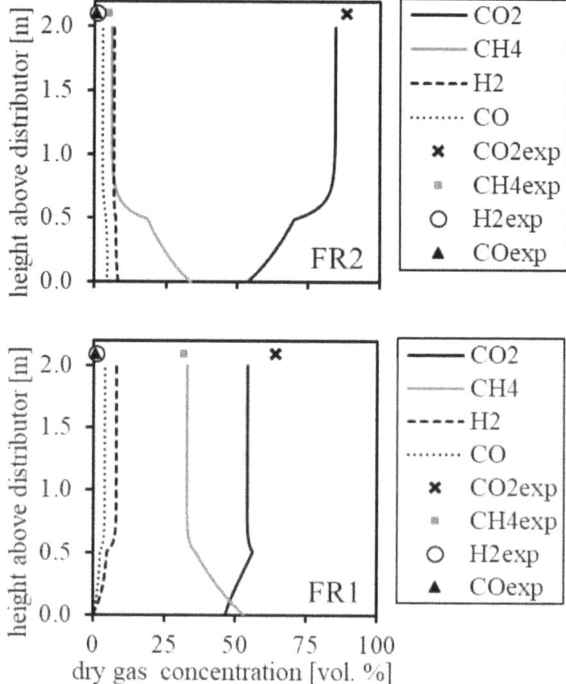


**Figure 65: Simulation of methane conversion inside the FR system with varied kinetic parameters. Shown concentrations in the lower dense zone are averaged over the bubble and suspension phase. Comparison to averaged gas concentration measurements.**

<u>Movement of bed masses inside the system</u>

**Experiment**

When fuel is added to the FR system, the volumetric gas flow inside is increasing, because one methane molecule is reacting to three product molecules. This will cause a higher superficial gas velocity in the reactor and, with it, the bubble volume fraction of the bed will increase [120]. The height of the bubbling bed is limited by the standpipe height, so in the end, less solids can be stored in the FR beds. This effect can be seen in the pressure drops during experiments, when methane is added, shown in Figure 66. The excess material from the FR moves to the AR and increases the solids holdup there. This effect is enlarged by the oxidation of the OC in the AR, which will reduce the volumetric gas flow there. When the fuel supply is stopped, the material moves back from the AR to the FR side. Due to the strong pressure fluctuations in the plant a precise determination of the solids holdup is difficult, but one can say that roughly 5-10% of the bed mass leave the FR system and on the other side the AR bed mass increases by 10-20%.

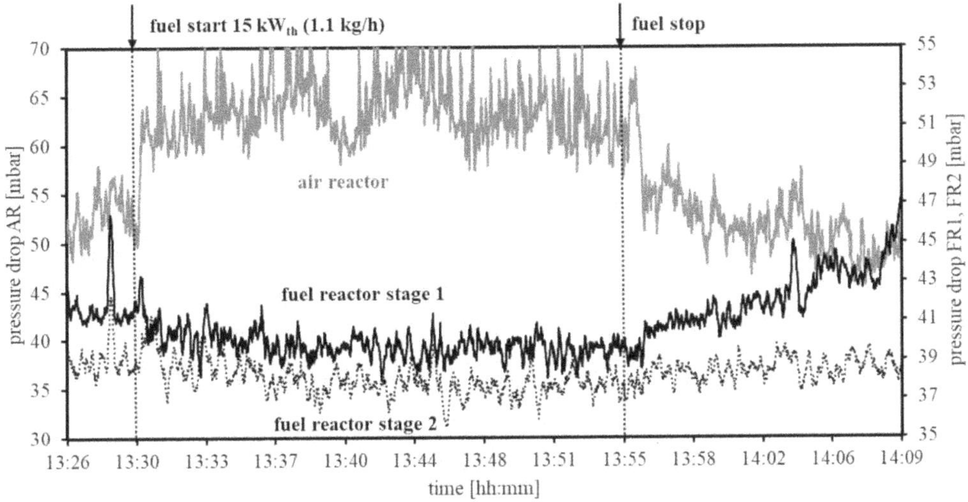


**Figure 66: Measured pressure drops over the whole reactor height of FR1, FR2 and AR. Methane injection started 13:30 and was stopped 13:55.**

## Simulation

The transient flowsheet simulation reveals the same behavior of the system that was observed in the experiments. Before the fuel is injected, all bed masses in the system plotted in Figure 67 arrange themselves in a stable operation point within 10-20 s from an arbitrary starting point. When methane injection starts at 100 s, bed material moves from both FR stages towards the AR. This is because of a higher gas volume flow, $CH_4$ reacts towards $CO_2$ and two molecules of $H_2O$, which increases the gas velocity inside both reactor stages. Stopping the fuel injection has the directly opposite effect and the bed masses arrange themselves basically the same way before fuel was inserted.

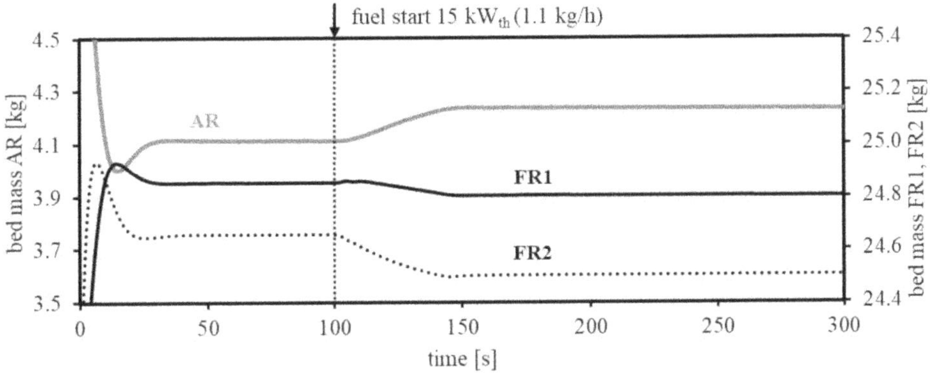


**Figure 67: Bed masses inside the system over time with a response to fuel injection at t = 100 s.**

Comparison of experiments and simulations shows that simulations can predict the response of the reactor system with regarding the bed mass distribution. In the simulations, the AR increases its bed mass by around 20%, when fuel is added. The FR loses around 5% of its bed mass, which is in the range of the experimental results.

Conversion of the oxygen carrier

The injection of methane converts the OC over a certain period starting from a state of total oxidation $X_s = 0$, which is plotted in Figure 68. The solid circulation rate $G_s$ was determined in previous experiments for the reactor system and was found to be within 15 $\frac{kg}{m^2\,s}$ and 45 $\frac{kg}{m^2\,s}$. The results of the chemical conversion of the OC are very sensitive to the solid circulation, because the oxygen carrier leaves the AR nearly fully oxidized. The more fully oxidized material enters the FR system, the lower is the overall OC conversion. The simulation environment predicts a solid circulation of 25 $\frac{kg}{m^2\,s}$with the settings from Table 3. After certain time the experimental points reveal that the reaction on the fuel start is covered correctly by the simulation environment. The fuel stop is a faster process in reality and the conversion of the OC samples drops faster than the simulation environment predicts.

With the help of the dynamic flowsheet simulation the solid conversion can be tracked and compared to the OC samples taken after 120, 240, 480 and 720 seconds of fuel injection, which is shown in Figure 68. Due to the big solid holdups inside the system and, further, the circulation of OC between the reactors, it takes more than 800 s in the simulation to reach a steady state after injection starts. The final conversion $X_s$ of the OC inside FR stage 1 is around 0.28. Stopping the fuel supply leads to the opposite effect and the conversion $X_s$ of the OC drops again towards 0. Comparing the simulation with the experimental observation, it becomes clear that the simulation captures the fuel start accurately, but in the simulation the reactor system takes considerably longer to go back to $X_s = 0$ after the fuel was stopped. This inaccuracy can be traced back to the assumed mixing behavior of the solids in the lower loop seal. There, a perfect mixing of solids hold-up with the incoming solids from the FR is assumed. In the pilot plant, it is probable that the solids move through like a moving bed without too much mixing of solids.

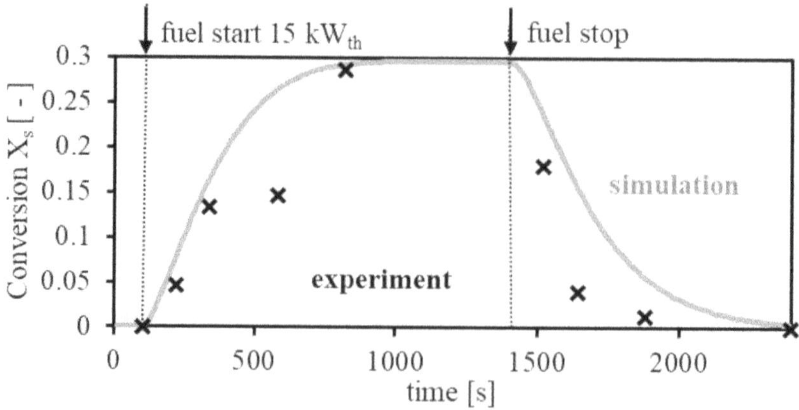


**Figure 68: Conversion of the OC from the lower stage FR1. Simulation of 2400 s of operation, fuel injection starts at 100 s and stops at 1400 s. Samples taken during hot operation after 120 s, 240 s, 480 s and 720 s after starting fuel injection and 120 s, 240 s,**

The re-oxidation of the depleted OC utilizes the oxygen from the air inside the AR. Due to the fast oxidization reaction of the OC, the amount of oxygen consumed is directly proportional to the mass flow and conversion state of the OC coming from the FR. The experimentally found oxygen concentrations after starting the fuel are plotted together with the results from the modeling in Figure 69. It is visible that the dynamic tracking of the OC conversion $X_s$ leads to an accurate prediction of the gas concentration prevailing in the AR over time. This means that the secondary particle property, the conversion state $X_s$, is correctly transferred from the fuel reactors and the AR.

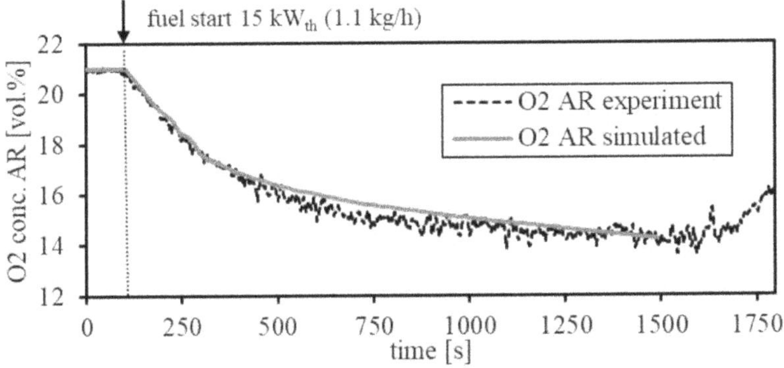


**Figure 69: Oxygen concentration at the AR outlet. Simulation of the system with fitted kinetics compared to gas concentrations found during the start of fuel injection.**

# 5. Summary and Conclusions

The overall goal of the work was the dynamic simulation of a system of interconnected fluidized bed reactors on the exemplary process Chemical Looping Combustion. The main challenges arose from the involved reactive solids and their complex reaction network as well as the influence of the main fluidized bed reactors on each other. This is why kinetic measurements were conducted in a lab scale fluidized bed and additionally pilot scale trials in a broad spectrum of operation parameters.

## 5.1. Experimental pilot plant

An experimental pilot plant for Chemical Looping Combustion was operated at 41 different operation conditions. The fuel type, methane, biomass, lignite and bituminous coal as well as temperature and fuel mass flow were varied. Figure 70 shows the range in which the parameters carbon capture and oxygen demand were found during the campaigns with lower stage fuel injection. The two-stage concept was able to convert fuel gases to a very high degree for all solid fuels and, hence, oxygen demand was below 3%. The gaseous fuel methane showed oxygen demand at 11%, which was traced back to the bypass of the gas in bubbles.

Carbon capture was excellent, i.e. over 95%, for lignite, methane and biomass. The char structure of bituminous coal led to higher gasification times and, with it, to a recognizable lower carbon capture in the system. When solid fuel was injected into the upper stage, higher oxygen demands, but also higher carbon capture for bituminous coal was found. This means that the system could be used with lower stage injection for the high volatile fuels and the upper stage injection for fuels with high carbon content.

Using lignite and bituminous coal, the fuel particle size had an impact on carbon capture. The bigger the fuel size, the more carbon was slipping to the air reactor. Biomass did not show this effect as it was basically gasified instantaneously regardless the fuel size. Compared to other one-stage designs it can be concluded that the two-stage design is well suited to reduce the oxygen demand, which is a major issue for in-site Gasification Chemical Looping Combustion.

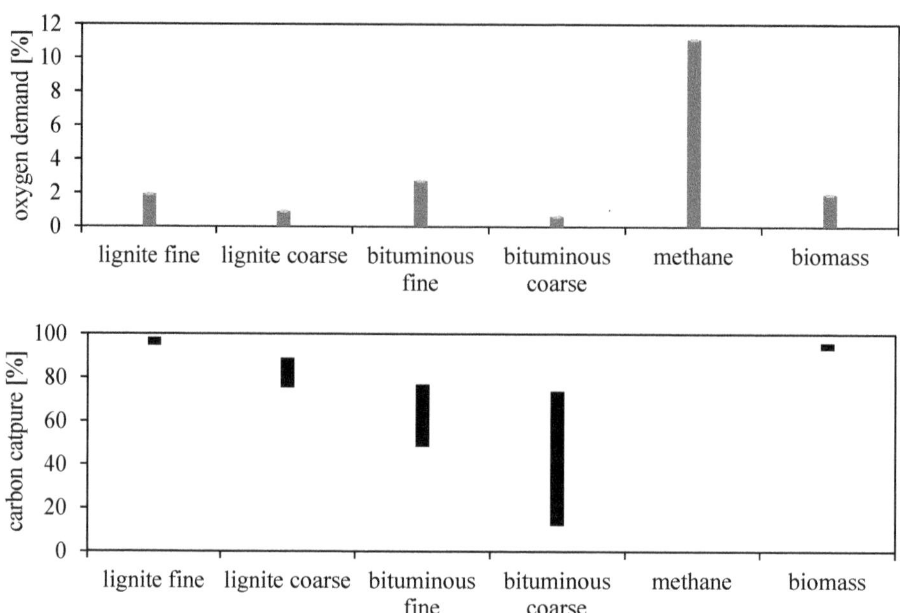


**Figure 70: Performance parameters ranges for oxygen demand and carbon capture of all pilot plant experiments where fuel was injected into the lower stage.**

The dynamics in the system were investigated by applying fuel starts and stops as well as load changes to the pilot plant system. When fuel injection started, the system took around 10-12 minutes to find new steady state operation conditions. A fuel stop was taking a bit shorter time for the system to reach a new equilibrium. When load changes were applied, similar response times of 10-12 minutes were found. Increasing the fuel flow from an already high fuel load, the response time rose significantly to around 18 minutes. It can be concluded that a complex char hold-up equilibrium with the oxidized oxygen carrier is formed.

## 5.2. Laboratory kinetic reactor

Starting from char gasification experiments in an inert bed with steam and $CO_2$, an analysis of the main mechanisms governing the conversion in in-situ Gasification Chemical Looping Combustion operation was conducted. A total of 169 batches of 1 g of lignite char were gasified in a laboratory scale fluidized bed reactor, whereof 135 runs were deemed successful with a closure of the carbon balance within a range of ± 25%. With a resistance approach for chemical reactions and mass transfer effects, kinetic parameters for the gasification are determined.

In an inert sand bed, the char gasification with steam is somewhat faster than with $CO_2$ as gasification agent. Simultaneous gasification with both agents results in gasification rates close to the one measured with steam only. Adding CO and $H_2$ to the fluidization gas using the gasification agent $CO_2$, the reaction rates drop by 50–60%. This shows the restraining effect by those two products of the gasification reactions. When a reactive oxygen carrier bed with fully oxidized oxygen carrier is present in the reactor, both steam and $CO_2$ gasification are accelerated compared to the inert bed case. Steam gasification is enhanced tenfold for the smallest particle size fraction; $CO_2$ gasification is enhanced six times compared to inert bed operation. These effects may be attributed to instantaneous consumption of the reaction products CO and $H_2$ by reaction with the oxygen carrier. Using coarser particles (> 1 mm) decreases the effect of the oxygen carrier presence and, hence, the gasification rates are only two to three times higher, which indicates a limitation of the rate by diffusion.

This diffusion limitation can be well captured with a resistance approach combining the limitations of chemical reactions and diffusion. Applying both $CO_2$ and steam simultaneously as char gasification agents is not enhancing the gasification rate further compared to steam-only gasification. This is valid for both inert bed and reactive oxygen carrier bed experiments. A set of kinetic parameters for the modeling of the char gasification based on the Shrinking Core Model are established from the experiments. For the reactive bed simultaneous $CO_2$ and steam gasification, a 0-order reaction with an activation energy of $E_a = 176$ kJ/mol was found. A pre-exponential factor of $k_0 = 2.2 \cdot 10^6$ holds. With the help of these parameters, the gasification rates of lignite char in a bed of reactive CuO-based oxygen carrier particles can be simulated in operation conditions between 800 °C and 1000 °C.

## 5.3. Dynamic Flowsheet Simulation

Dynamic unit models needed for CLC with methane were developed and implemented into the novel Dyssol simulation software. The developed unit models for CLC comprise a CFB riser reactor, a bubbling bed reactor, a cyclone and the loop seal. The dynamic flowsheet models were used to simulate the hot operation of an actual CLC plant with methane used as a fuel. The operation was simulated over a duration of 40 min. Within this timeframe the fuel supply was started and stopped again after several minutes. The simulation results were compared to the experimental data from the pilot plant, where the same dynamic changes were applied.

Starting the fuel injection inside the experimental plant made the bed masses rearrange inside the system within around 30 s, which could be reproduced with the dynamic flowsheet simulation. It took around 15 min to find a steady state for the solid conversion of the OC after fuel injection started. In a first simulation approach, using literature kinetics data for the used OC from TGA measurement, it turned out that the $CH_4$ conversion was overpredicted and CO and $H_2$ conversion were underpredicted. The reasons for these deviations between simulation and measurements may either be due to differences in the chemical structure and composition of the OCs or due to different mass and heat transfer conditions in TGA and fluidized bed reactor, respectively. A second simulation with fitted kinetic parameters gave a good description of the measurements. It was possible to correctly describe the transient OC conversion inside the interconnected fluidized beds after a start of fuel feeding.

Computational costs of the used flowsheet simulation are low compared to CFD simulations what makes real-time simulations of industrial scale combustion systems feasible. These real-time simulations can later be implemented into operation control and predict the behavior of whole plants at load-changes what still is a major issue during operation.

# List of Abbreviations

Abbreviations, which were only used once and explained in text are not listed.

| **Abbreviation** | **Explanation** |
|---|---|
| AR | Air reactor |
| BECCS | Bio-energy carbon capture and storage |
| BFB | Bubbling fluidized bed |
| CCS | Carbon capture and storage |
| CFB | Circulating fluidized bed reactor |
| CFD | Computational fluid dynamics |
| CLC | Chemical looping combustion |
| CLOU | Chemical looping with oxygen uncoupling |
| CLR | Chemical looping reforming |
| CSTR | Continuously stirred reactor |
| FR | Fuel reactor |
| iG-CLC | in-situ Gasification CLC |
| OC | Oxygen carrier |
| PFR | Plug flow reactor |
| PSD | Particle size distribution |
| SC | Supply chamber |
| RC | Recycle chamber |
| SCM | Shrinking core model |
| TGA | Thermogravimetric analyzer |
| TUHH | Hamburg University of Technology |
| VRM | Volumetric reaction model |
| WGS | Water-gas shift reaction |

# List of Symbols

| Symbol/letter | Description | Unit |
|---|---|---|
| $j_{Q,l}$ | Convective flow between suspension and bubble phase | [mol/m$^3$] |
| $\dot{m}_i$ | Mass flow out of plant component i | [kg/s] |
| $\dot{n}_i$ | molar flow of component i | [mol/s] |
| $\dot{V}_b$ | Visible bubble flow | [m/s] |
| $\dot{V}_{or}$ | Volumetric flow rate through orifice | [m$^3$/s] |
| a | Decay constant | [1/m] |
| $A_r$ | Fluidized bed reactor cross section area | [m$^2$] |
| $a_t$ | Exchange area between bubbles and suspension | [1/m] |
| $C_{b,l}$ | Gas concentration of gas component l in bubble phase | [mol/m$^3$] |
| CC | Conversion class size | [%/total] |
| $C_{d,l}$ | Gas concentration of gas component l in dense phase | [mol/m$^3$] |
| $C_{dist}$ | Conversion distribution | [%] |
| $C_{f,l}$ | Gas concentration of gas component l in the freeboard | [mol/m$^3$] |
| $C_l$ | Gas concentration of gas component l | [mol/m$^3$] |
| $c_v$ | Volumetric solids concentration | [ - ] |
| $c_{v\_mf}$ | Solids concentration at minimum fluidization velocity | [ - ] |
| $c_{v\_suspension}$ | Solids concentration in the dense suspension phase | [ - ] |
| D | Molar binary diffusion coefficient | [m$^2$/s] |
| $d_{i,p}$ | Average diameter of size interval i | [m] |
| $d_v$ | Bubble size | [m] |
| $d_{v,0}$ | Initial bubble size | [m] |
| $E_A$ | Activation energy | [J/mol] |
| $F_i$ | Volumetric flow of gas component i | [m$^3$/s] |
| g | Gravitational acceleration | [m/s$^2$] |
| $G_{s,i,\infty}$ | Solids circulation rate of size interval i | [kg/m$^2$/s] |
| h | Actual height inside reactor | [m] |
| $H_b$ | End of dense bottom zone, height above distributor | [m] |
| k | Kinetic constant for certain temperature and concentration | [1/s] |
| $k_0$ | Pre-exponential factor | [1/s] |
| $k_g$ | Gas diffusion resistance | [mol/s] |
| $K_{i,\infty}$ | Entrainment constant for each particle size interval i | [kg/m$^2$/s] |
| $K_Q$ | Convective exchange rate | [1/s] |
| $m_r$ | Mass inside reactor | [kg] |
| n | Reaction order | [ - ] |
| $Q_3$ | Cumulative mass-based particle size distribution | [%] |

| | | |
|---|---|---|
| $R_g$ | Rate of gasification | [1/s] |
| rg,l | reaction rate of gas l | [mol/m$^3$] |
| $r_{s,j,l}$ | Reaction rate of solid component j towards gas l | [mol/m$^3$/s] |
| $T(u_{t,i})$ | Separation efficiency curve | [ - ] |
| $u_b$ | Bubble rise velocity | [m/s] |
| $u_R$ | Superficial gas velocity | [m/s] |
| $u_{t,i}$ | Terminal velocity of particles in size interval i | [m/s] |
| $w_C$ | Mass of char gasified | [kg] |
| $x_i$ | Molar fraction of gas component i | [ - ] |
| $X_S$ | Solid conversion | [ - ] |
| $X_{SF}$ | Solid fuel conversion | [ - ] |
| $\varepsilon_b$ | Bubble volume fraction | [–] |
| ηCC | Carbon capture efficiency | [ - ] |
| ηf | Cyclone separation efficiency | [ - ] |
| ηOO | Oxide oxygen efficiency | [ - ] |
| θ | Scale dependent geometry parameters | [ - ] |
| λ | Bubble lifetime | [s] |
| $\rho_f$ | Density fluid | [kg/m$^3$] |
| $\rho_m$ | Molar density | [mol/m$^3$] |
| $\rho_S$ | Density solid | [kg/m$^3$] |
| $\Omega_{OD}$ | Oxygen demand | [ - ] |

## References


[1] United Nations Framework Convention on Climate Change, Paris Agreement: UNFCCC, 2015.

[2] G.P. Peters, The 'best available science' to inform 1.5 °C policy choices, Nature Clim Change 6 (2016) 646–649. https://doi.org/10.1038/nclimate3000.

[3] P. Smith, S.J. Davis, F. Creutzig, S. Fuss, J. Minx, B. Gabrielle, E. Kato, R.B. Jackson, A. Cowie, E. Kriegler, D.P. van Vuuren, J. Rogelj, P. Ciais, J. Milne, J.G. Canadell, D. McCollum, G. Peters, R. Andrew, V. Krey, G. Shrestha, P. Friedlingstein, T. Gasser, A. Grübler, W.K. Heidug, M. Jonas, C.D. Jones, F. Kraxner, E. Littleton, J. Lowe, J.R. Moreira, N. Nakicenovic, M. Obersteiner, A. Patwardhan, M. Rogner, E. Rubin, A. Sharifi, A. Torvanger, Y. Yamagata, J. Edmonds, C. Yongsung, Biophysical and economic limits to negative CO2 emissions, Nature Clim Change 6 (2016) 42–50. https://doi.org/10.1038/NCLIMATE2870.

[4] K. Anderson, G. Peters, The trouble with negative emissions, Science 354 (2016) 182–183. https://doi.org/10.1126/science.aah4567.

[5] T. Mendiara, P. Gayán, F. García-Labiano, L.F. de Diego, A. Pérez-Astray, M.T. Izquierdo, A. Abad, J. Adánez, Chemical Looping Combustion of Biomass: An Approach to BECCS, Energy Procedia 114 (2017) 6021–6029. https://doi.org/10.1016/j.egypro.2017.03.1737.

[6] A. Lyngfelt, B. Leckner, T. Mattisson, A fluidized-bed combustion process with inherent CO2 separation; application of chemical-looping combustion, Chemical Engineering Science 56 (2001) 3101–3113. https://doi.org/10.1016/S0009-2509(01)00007-0.

[7] J. Gartung, Absenkung der Mindestlast eines braunkohlebefeuerten Bestandskraftwerks, 1st ed., Cuvillier Verlag, Göttingen, 2018.

[8] O. Geden, G. Peters, V. Scott (Eds.), ‘Full’ vs. ‘limited CDR’ – how to get EU climate policymakers on board, 2018.

[9] A. Lyngfelt, T. Mattisson, M. Ryden, C. Linderholm (Eds.), 10,000 h of Chemical-Looping Combustion Operation: Where Are We and Where Do We Want to Go?, 2018.

[10] J. Adanez, A. Abad, F. Garcia-Labiano, P. Gayan, L.F. de Diego, Progress in Chemical-Looping Combustion and Reforming technologies, Progress in Energy and Combustion Science 38 (2012) 215–282. https://doi.org/10.1016/j.pecs.2011.09.001.

[11] J. Adánez, A. Abad, T. Mendiara, P. Gayán, L.F. de Diego, F. García-Labiano, Chemical looping combustion of solid fuels, Progress in Energy and Combustion Science 65 (2018) 6–66. https://doi.org/10.1016/j.pecs.2017.07.005.

[12] A. Lyngfelt, C. Linderholm, Chemical-Looping Combustion of Solid Fuels – Status and Recent Progress, Energy Procedia 114 (2017) 371–386. https://doi.org/10.1016/j.egypro.2017.03.1179.

[13] A. Lyngfelt, Chemical-looping combustion of solid fuels – Status of development, Applied Energy 113 (2014) 1869–1873. https://doi.org/10.1016/j.apenergy.2013.05.043.

[14] T. Song, L. Shen, Review of reactor for chemical looping combustion of solid fuels, International Journal of Greenhouse Gas Control 76 (2018) 92–110. https://doi.org/10.1016/j.ijggc.2018.06.004.

[15] A. Thon, Operation of a System of Interconnected Fluidized Bed Reactors in the Chemical Looping Combustion Process, 1st ed., Cuvillier, E, Göttingen, Niedersachs, 2014.

[16] M. Kramp, Chemical looping combustion in interconnected fluidized bed reactors: Simulation and experimental validation, 1st ed., Verl. Dr. Hut, München, 2014.

[17] A. Tong, S. Bayham, M.V. Kathe, L. Zeng, S. Luo, L.-S. Fan, Iron-based syngas chemical looping process and coal-direct chemical looping process development at Ohio State University, Applied Energy 113 (2014) 1836–1845. https://doi.org/10.1016/j.apenergy.2013.05.024.

[18] L. Zeng, A. Tong, M. Kathe, S. Bayham, L.-S. Fan, Iron oxide looping for natural gas conversion in a countercurrent moving bed reactor, Applied Energy 157 (2015) 338–347. https://doi.org/10.1016/j.apenergy.2015.06.029.

[19] J.M. Weber, R.C. Stehle, R.W. Breault, J. de Wilde, Experimental study of the application of rotating fluidized beds to particle separation, Powder Technology 316 (2017) 123–130. https://doi.org/10.1016/j.powtec.2016.12.076.

[20] D. Geldart, Types of gas fluidization, Powder Technology 7 (1973) 285–292. https://doi.org/10.1016/0032-5910(73)80037-3.

[21] K.-E. Wirth, Zirkulierende Wirbelschichten: Strömungsmechanische Grundlagen, Anwendung in der Feuerungstechnik, Springer Berlin Heidelberg, Berlin, Heidelberg, s.l., 1990.

[22] J. Werther, S. Heinrich, M. Dosta, E.-U. Hartge, The ultimate goal of modeling—Simulation of system and plant performance, Particuology 9 (2011) 320–329. https://doi.org/10.1016/j.partic.2011.03.006.

[23] A. Abad, J. Adánez, L.F. de Diego, P. Gayán, F. García-Labiano, A. Lyngfelt, Fuel reactor model validation: Assessment of the key parameters affecting the chemical-looping combustion of coal, International Journal of Greenhouse Gas Control 19 (2013) 541–551. https://doi.org/10.1016/j.ijggc.2013.10.020.

[24] A. Puettmann, E.-U. Hartge, J. Werther, Application of the flowsheet simulation concept to fluidized bed reactor modeling. Part II—Application to the selective oxidation of n-butane to maleic anhydride in a riser/regenerator system, Chemical Engineering and Processing: Process Intensification 57-58 (2012) 86–95. https://doi.org/10.1016/j.cep.2011.10.004.

[25] D. Kunii, O. Levenspiel, H. Brenner, Fluidization Engineering, 2nd ed., Elsevier Science, Saint Louis, 2014.

[26] W.-C. Yang, Handbook of fluidization and fluid-particle systems, Dekker, New York, 2003.

[27] J.R. Grace, J.M. Matsen, Fluidization, Springer US, Boston, MA, 1980.

[28] P. Basu, M. Horio, M. Hasatani, Circulating fluidized bed technology III: Proceedings of the Third International Conference on Circulating Fluidized Beds, Nagoya, Japan, 14-18 October 1990, 1st ed., Pergamon Press, Oxford, New York, 1991.

[29] D. Bai, K. Kato, Quantitative estimation of solids holdups at dense and dilute regions of circulating fluidized beds, Powder Technology 101 (1999) 183–190. https://doi.org/10.1016/S0032-5910(98)00159-4.

[30] J. Werther, E.-U. Hartge, A population balance model of the particle inventory in a fluidized-bed reactor/regenerator system, Powder Technology 148 (2004) 113–122. https://doi.org/10.1016/j.powtec.2004.09.005.

[31] J. Werther, C. Aue-Klett, A. Al-Shawabkeh (Eds.), Gas and solids mixing in the bottom zone of a circulating fluidized bed, 2002.

[32] G.S. Patience, J. Chaouki, Gas phase hydrodynamics in the riser of a circulating fluidized bed, Chemical Engineering Science 48 (1993) 3195–3205. https://doi.org/10.1016/0009-2509(93)80205-5.

[33] A.E. Fouda, C.E. Capes, Calculation of large numbers of terminal velocities or equivalent particle diameters using polynomial equations fitted to the Heywood tables, Powder Technology 13 (1976) 291–293. https://doi.org/10.1016/0032-5910(76)85016-4.

[34] A.S. Issangya, J.R. Grace, D. Bai, J. Zhu, Radial voidage variation in CFB risers, Can. J. Chem. Eng. 79 (2001) 279–286. https://doi.org/10.1002/cjce.5450790211.
[35] W. Zhang, Y. Tung, F. Johnsson, Radial voidage profiles in fast fluidized beds of different diameters, Chemical Engineering Science 46 (1991) 3045–3052. https://doi.org/10.1016/0009-2509(91)85008-L.
[36] T.S. Pugsley, F. Berruti, A predictive hydrodynamic model for circulating fluidized bed risers, Powder Technology 89 (1996) 57–69. https://doi.org/10.1016/S0032-5910(96)03154-3.
[37] W.K. Lewis, E.R. Gilliland 2665971, 1954.
[38] M. Ishida, D. Zheng, T. Akehata, Evaluation of a chemical-looping-combustion power-generation system by graphic exergy analysis, Energy 12 (1987) 147–154. https://doi.org/10.1016/0360-5442(87)90119-8.
[39] M. Ishida, H. Jin, A new advanced power-generation system using chemical-looping combustion, Energy 19 (1994) 415–422. https://doi.org/10.1016/0360-5442(94)90120-1.
[40] M. Ishida, H. Jin, A Novel Chemical-Looping Combustor without NOx Formation, Ind. Eng. Chem. Res. 35 (1996) 2469–2472. https://doi.org/10.1021/ie950680s.
[41] Q. Song, R. Xiao, Z. Deng, L. Shen, J. Xiao, M. Zhang, Effect of Temperature on Reduction of CaSO 4 Oxygen Carrier in Chemical-Looping Combustion of Simulated Coal Gas in a Fluidized Bed Reactor, Ind. Eng. Chem. Res. 47 (2008) 8148–8159. https://doi.org/10.1021/ie8007264.
[42] T. Mattisson, A. Lyngfelt, H. Leion, Chemical-looping with oxygen uncoupling for combustion of solid fuels, International Journal of Greenhouse Gas Control 3 (2009) 11–19. https://doi.org/10.1016/j.ijggc.2008.06.002.
[43] F. Scala (Ed.), Fluidized bed technologies for near-zero emission combustion and gasification, Woodhead Publishing, Oxford [etc.], op. 2013.
[44] D.M.B. Adams, J. Davison, Capturing $CO_2$, IEA Greenhouse Gas R & D Programme, [Cheltenham, Eng.?], 2007.
[45] A. Lyngfelt, Oxygen carriers for chemical-looping combustion, in: Calcium and Chemical Looping Technology for Power Generation and Carbon Dioxide (CO2) Capture, Elsevier, 2015, pp. 221–254.
[46] M. Ryden, A. Lyngfelt, Using steam reforming to produce hydrogen with carbon dioxide capture by chemical-looping combustion, International Journal of Hydrogen Energy 31 (2006) 1271–1283. https://doi.org/10.1016/j.ijhydene.2005.12.003.
[47] M. Ortiz, A. Abad, L.F. de Diego, F. García-Labiano, P. Gayán, J. Adánez, Optimization of hydrogen production by Chemical-Looping auto-thermal Reforming working with Ni-based oxygen-carriers, International Journal of Hydrogen Energy 36 (2011) 9663–9672. https://doi.org/10.1016/j.ijhydene.2011.05.025.
[48] P. Markström, C. Linderholm, A. Lyngfelt, Chemical-looping combustion of solid fuels – Design and operation of a 100kW unit with bituminous coal, International Journal of Greenhouse Gas Control 15 (2013) 150–162. https://doi.org/10.1016/j.ijggc.2013.01.048.
[49] C. Linderholm, A. Lyngfelt, A. Cuadrat, E. Jerndal, Chemical-looping combustion of solid fuels – Operation in a 10kW unit with two fuels, above-bed and in-bed fuel feed and two oxygen carriers, manganese ore and ilmenite, Fuel 102 (2012) 808–822. https://doi.org/10.1016/j.fuel.2012.05.010.
[50] C. Linderholm, M. Schmitz, Chemical-looping combustion of solid fuels in a 100 kW dual circulating fluidized bed system using iron ore as oxygen carrier, Journal of Environmental Chemical Engineering 4 (2016) 1029–1039. https://doi.org/10.1016/j.jece.2016.01.006.

[51] C. Linderholm, M. Schmitz, P. Knutsson, A. Lyngfelt, Chemical-looping combustion in a 100-kW unit using a mixture of ilmenite and manganese ore as oxygen carrier, Fuel 166 (2016) 533–542. https://doi.org/10.1016/j.fuel.2015.11.015.

[52] P. Markström, C. Linderholm, A. Lyngfelt, Operation of a 100kW chemical-looping combustor with Mexican petroleum coke and Cerrejón coal, Applied Energy 113 (2014) 1830–1835. https://doi.org/10.1016/j.apenergy.2013.04.066.

[53] M. Schmitz, C.J. Linderholm, Performance of calcium manganate as oxygen carrier in chemical looping combustion of biochar in a 10kW pilot, Applied Energy 169 (2016) 729–737. https://doi.org/10.1016/j.apenergy.2016.02.088.

[54] T. Mendiara, A. Abad, L.F. de Diego, F. García-Labiano, P. Gayán, J. Adánez, Biomass combustion in a CLC system using an iron ore as an oxygen carrier, International Journal of Greenhouse Gas Control 19 (2013) 322–330. https://doi.org/10.1016/j.ijggc.2013.09.012.

[55] I. Adánez-Rubio, A. Abad, P. Gayán, L.F. de Diego, F. García-Labiano, J. Adánez, Biomass combustion with CO2 capture by chemical looping with oxygen uncoupling (CLOU), Fuel Processing Technology 124 (2014) 104–114. https://doi.org/10.1016/j.fuproc.2014.02.019.

[56] T. Mendiara, A. Pérez-Astray, M.T. Izquierdo, A. Abad, L.F. de Diego, F. García-Labiano, P. Gayán, J. Adánez, Chemical Looping Combustion of different types of biomass in a 0.5 kW th unit, Fuel 211 (2018) 868–875. https://doi.org/10.1016/j.fuel.2017.09.113.

[57] R. Pérez-Vega, A. Abad, F. García-Labiano, P. Gayán, L.F. de Diego, J. Adánez, Coal combustion in a 50kWth Chemical Looping Combustion unit: Seeking operating conditions to maximize CO2 capture and combustion efficiency, International Journal of Greenhouse Gas Control 50 (2016) 80–92. https://doi.org/10.1016/j.ijggc.2016.04.006.

[58] A. Abad, I. Adánez-Rubio, P. Gayán, F. García-Labiano, L.F. de Diego, J. Adánez, Demonstration of chemical-looping with oxygen uncoupling (CLOU) process in a 1.5kWth continuously operating unit using a Cu-based oxygen-carrier, International Journal of Greenhouse Gas Control 6 (2012) 189–200. https://doi.org/10.1016/j.ijggc.2011.10.016.

[59] A. Abad, R. Pérez-Vega, L.F. de Diego, F. García-Labiano, P. Gayán, J. Adánez, Design and operation of a 50 kW th Chemical Looping Combustion (CLC) unit for solid fuels, Applied Energy 157 (2015) 295–303. https://doi.org/10.1016/j.apenergy.2015.03.094.

[60] I. Adánez-Rubio, A. Abad, P. Gayán, L.F. de Diego, F. García-Labiano, J. Adánez, Performance of CLOU process in the combustion of different types of coal with CO2 capture, International Journal of Greenhouse Gas Control 12 (2013) 430–440. https://doi.org/10.1016/j.ijggc.2012.11.025.

[61] X. Wang, B. Jin, X. Zhu, H. Liu, Experimental Evaluation of a Novel 20 kW th in Situ Gasification Chemical Looping Combustion Unit with an Iron Ore as the Oxygen Carrier, Ind. Eng. Chem. Res. 55 (2016) 11775–11784. https://doi.org/10.1021/acs.iecr.6b03028.

[62] H. Gu, L. Shen, Z. Zhong, X. Niu, H. Ge, Y. Zhou, S. Xiao, Potassium-Modified Iron Ore as Oxygen Carrier for Coal Chemical Looping Combustion: Continuous Test in 1 kW Reactor, Ind. Eng. Chem. Res. 53 (2014) 13006–13015. https://doi.org/10.1021/ie501328h.

[63] J. Ma, H. Zhao, X. Tian, Y. Wei, S. Rajendran, Y. Zhang, S. Bhattacharya, C. Zheng, Chemical looping combustion of coal in a 5 kW th interconnected fluidized bed reactor using hematite as oxygen carrier, Applied Energy 157 (2015) 304–313. https://doi.org/10.1016/j.apenergy.2015.03.124.

[64] P. Ohlemüller, J.-P. Busch, M. Reitz, J. Ströhle, B. Epple, Chemical-Looping Combustion of Hard Coal: Autothermal Operation of a 1 MW th Pilot Plant, J. Energy Resour. Technol 138 (2016) 42203. https://doi.org/10.1115/1.4032357.

[65] J. Ströhle, M. Orth, B. Epple, Design and operation of a 1MWth chemical looping plant, Applied Energy 113 (2014) 1490–1495. https://doi.org/10.1016/j.apenergy.2013.09.008.

[66] S. Bayham, O. McGiveron, A. Tong, E. Chung, M. Kathe, D. Wang, L. Zeng, L.-S. Fan, Parametric and dynamic studies of an iron-based 25-kW th coal direct chemical looping unit using sub-bituminous coal, Applied Energy 145 (2015) 354–363. https://doi.org/10.1016/j.apenergy.2015.02.026.

[67] T. Pikkarainen, I. Hiltunen, S. Teir (Eds.), Piloting of bio-CLC for BECCS, 2016.

[68] O. Levenspiel, Chemical reaction engineering, 3rd ed., Wiley, Hoboken, NJ, 1999.

[69] F. García-Labiano, L.F. de Diego, J. Adánez, A. Abad, P. Gayán, Reduction and Oxidation Kinetics of a Copper-Based Oxygen Carrier Prepared by Impregnation for Chemical-Looping Combustion, Ind. Eng. Chem. Res. 43 (2004) 8168–8177. https://doi.org/10.1021/ie0493311.

[70] A. Abad, F. García-Labiano, L.F. de Diego, P. Gayán, J. Adánez, Reduction Kinetics of Cu-, Ni-, and Fe-Based Oxygen Carriers Using Syngas (CO + H 2 ) for Chemical-Looping Combustion, Energy Fuels 21 (2007) 1843–1853. https://doi.org/10.1021/ef070025k.

[71] W. Hu, F. Donat, S.A. Scott, J.S. Dennis, Kinetics of oxygen uncoupling of a copper based oxygen carrier, Applied Energy 161 (2016) 92–100. https://doi.org/10.1016/j.apenergy.2015.10.006.

[72] S.Y. Chuang, J.S. Dennis, A.N. Hayhurst, S.A. Scott, Kinetics of the chemical looping oxidation of H2 by a co-precipitated mixture of CuO and Al2O3, Chemical Engineering Research and Design 89 (2011) 1511–1523. https://doi.org/10.1016/j.cherd.2010.08.003.

[73] H. Kruggel-Emden, F. Stepanek, A. Munjiza, A comparative study of reaction models applied for chemical looping combustion, Chemical Engineering Research and Design 89 (2011) 2714–2727. https://doi.org/10.1016/j.cherd.2011.05.013.

[74] Thermal methods of analysis: Principles, applications and problems, Springer-Science+Business Media, B.V, Dordrecht, 1995.

[75] J.M. Berty, Testing Commercial Catalysts in Recycle Reactors, Catalysis Reviews 20 (1979) 75–96. https://doi.org/10.1080/03602457908065106.

[76] Werner Sitzmann, Experimentelle Untersuchung und mathematische Modellierung komplexer heterogenkatalytischer Reaktionen in Wirbelschichten am Beispiel der Ethanoldehydratisierung, 1986.

[77] J. Yu, C. Yao, X. Zeng, S. Geng, L. Dong, Y. Wang, S. Gao, G. Xu, Biomass pyrolysis in a micro-fluidized bed reactor: Characterization and kinetics, Chemical Engineering Journal 168 (2011) 839–847. https://doi.org/10.1016/j.cej.2011.01.097.

[78] A. Abad, J. Adánez, F. García-Labiano, L.F. de Diego, P. Gayán, Modeling of the chemical-looping combustion of methane using a Cu-based oxygen-carrier, Combustion and Flame 157 (2010) 602–615. https://doi.org/10.1016/j.combustflame.2009.10.010.

[79] L. Lindmüller, Kinetics and Fluid Mechanics of Oxygen Carriers for Chemical Looping Combustion. Master Thesis, Hamburg, Germany, 2018.

[80] S.Y. Chuang, J.S. Dennis, A.N. Hayhurst, S.A. Scott, Kinetics of the chemical looping oxidation of CO by a co-precipitated mixture of CuO and Al2O3, Proceedings of the Combustion Institute 32 (2009) 2633–2640. https://doi.org/10.1016/j.proci.2008.06.112.

[81] S.Y. Chuang, J.S. Dennis, A.N. Hayhurst, S.A. Scott, Kinetics of the Oxidation of a Co-precipitated Mixture of Cu and Al 2 O 3 by O 2 for Chemical-Looping Combustion, Energy Fuels 24 (2010) 3917–3927. https://doi.org/10.1021/ef1002167.

[82] C.D. Bohn, J.P. Cleeton, C.R. Müller, J.F. Davidson, A.N. Hayhurst, S.A. Scott, J.S. Dennis, The kinetics of the reduction of iron oxide by carbon monoxide mixed with carbon dioxide, AIChE J. 47 (2010) NA-NA. https://doi.org/10.1002/aic.12084.

[83] E. Ksepko, M. Sciazko, P. Babinski, Studies on the redox reaction kinetics of Fe2O3–CuO/Al2O3 and Fe2O3/TiO2 oxygen carriers, Applied Energy 115 (2014) 374–383. https://doi.org/10.1016/j.apenergy.2013.10.064.

[84] Z. Zhou, L. Han, O. Nordness, G.M. Bollas, Continuous regime of chemical-looping combustion (CLC) and chemical-looping with oxygen uncoupling (CLOU) reactivity of CuO oxygen carriers, Applied Catalysis B: Environmental 166-167 (2015) 132–144. https://doi.org/10.1016/j.apcatb.2014.10.067.

[85] P. Gayán, C.R. Forero, A. Abad, L.F. de Diego, F. García-Labiano, J. Adánez, Effect of Support on the Behavior of Cu-Based Oxygen Carriers during Long-Term CLC Operation at Temperatures above 1073 K, Energy Fuels 25 (2011) 1316–1326. https://doi.org/10.1021/ef101583w.

[86] P.H. Bolt, F.H.P.M. Habraken, J.W. Geus, Formation of Nickel, Cobalt, Copper, and Iron Aluminates fromα- andγ-Alumina-Supported Oxides: A Comparative Study, Journal of Solid State Chemistry 135 (1998) 59–69. https://doi.org/10.1006/jssc.1997.7590.

[87] F. Winkler (IG FARBENINDUSTRIE AG) DE437970 (C), 1922.

[88] J.L. Johnson, Kinetics of coal gasification: A compilation of research, Wiley, New York, 1979.

[89] I. Matsui, T. Kojima, D. Kunii, T. Furusawa, Study of char gasification by carbon dioxide. 2. Continuous gasification in fluidized bed, Ind. Eng. Chem. Res. 26 (1987) 95–100. https://doi.org/10.1021/ie00061a018.

[90] I. Matsui, D. Kunii, T. Furusawa, Study of fluidized bed steam gasification of char by thermogravimetrically obtained kinetics, J. Chem. Eng. Japan / JCEJ 18 (1985) 105–113. https://doi.org/10.1252/jcej.18.105.

[91] I. Matsui, D. Kunii, T. Furusawa, Study of char gasification by carbon dioxide. 1. Kinetic study by thermogravimetric analysis, Ind. Eng. Chem. Res. 26 (1987) 91–95. https://doi.org/10.1021/ie00061a017.

[92] M. Troiano, P. Ammendola, F. Scala, Attrition of lignite char under fluidized bed gasification conditions: The effect of carbon conversion, Proceedings of the Combustion Institute 34 (2013) 2741–2747. https://doi.org/10.1016/j.proci.2012.05.105.

[93] A. Cuadrat, A. Abad, L.F. de Diego, F. García-Labiano, P. Gayán, J. Adánez, Prompt considerations on the design of Chemical-Looping Combustion of coal from experimental tests, Fuel 97 (2012) 219–232. https://doi.org/10.1016/j.fuel.2012.01.050.

[94] A. Cuadrat, A. Abad, P. Gayán, L.F. de Diego, F. García-Labiano, J. Adánez, Theoretical approach on the CLC performance with solid fuels: Optimizing the solids inventory, Fuel 97 (2012) 536–551. https://doi.org/10.1016/j.fuel.2012.01.071.

[95] S.A. Scott, J.S. Dennis, A.N. Hayhurst, T. Brown, In situ gasification of a solid fuel and CO 2 separation using chemical looping, AIChE J. 52 (2006) 3325–3328. https://doi.org/10.1002/aic.10942.

[96] T.A. Brown, J.S. Dennis, S.A. Scott, J.F. Davidson, A.N. Hayhurst, Gasification and Chemical-Looping Combustion of a Lignite Char in a Fluidized Bed of Iron Oxide, Energy Fuels 24 (2010) 3034–3048. https://doi.org/10.1021/ef100068m.

[97] M. Keller, H. Leion, T. Mattisson, A. Lyngfelt, Gasification inhibition in chemical-looping combustion with solid fuels, Combustion and Flame 158 (2011) 393–400. https://doi.org/10.1016/j.combustflame.2010.09.009.

[98] M. Arjmand, H. Leion, T. Mattisson, A. Lyngfelt, Investigation of different manganese ores as oxygen carriers in chemical-looping combustion (CLC) for solid fuels, Applied Energy 113 (2014) 1883–1894. https://doi.org/10.1016/j.apenergy.2013.06.015.

[99] J. Bolhàr-Nordenkampf, T. Pröll, P. Kolbitsch, H. Hofbauer, Comprehensive Modeling Tool for Chemical Looping Based Processes, Chem. Eng. Technol. 32 (2009) 410–417. https://doi.org/10.1002/ceat.200800568.

[100] P. Peltola, J. Ritvanen, T. Tynjälä, T. Hyppänen, Fuel reactor modelling in chemical looping with oxygen uncoupling process, Fuel 147 (2015) 184–194. https://doi.org/10.1016/j.fuel.2015.01.073.

[101] P. Peltola, J. Ritvanen, T. Tynjälä, T. Pröll, T. Hyppänen, One-dimensional modelling of chemical looping combustion in dual fluidized bed reactor system, International Journal of Greenhouse Gas Control 16 (2013) 72–82. https://doi.org/10.1016/j.ijggc.2013.03.008.

[102] R. Porrazzo, G. White, R. Ocone, Aspen Plus simulations of fluidised beds for chemical looping combustion, Fuel 136 (2014) 46–56. https://doi.org/10.1016/j.fuel.2014.06.053.

[103] S. Mukherjee, P. Kumar, A. Yang, P. Fennell, A systematic investigation of the performance of copper-, cobalt-, iron-, manganese- and nickel-based oxygen carriers for chemical looping combustion technology through simulation models, Chemical Engineering Science 130 (2015) 79–91. https://doi.org/10.1016/j.ces.2015.03.009.

[104] F. Li, L. Zeng, D. Sridhar, L.G. Velazquez-Vargas, L.-S. Fan, Chemical Looping Gasification Using Solid Fuels, in: L.-S. Fan (Ed.), Chemical Looping Systems for Fossil Energy Conversions, John Wiley & Sons, Inc, Hoboken, NJ, USA, 2010, pp. 301–361.

[105] A.H. Sahir, J.K. Dansie, A.L. Cadore, J.S. Lighty, A comparative process study of chemical-looping combustion (CLC) and chemical-looping with oxygen uncoupling (CLOU) for solid fuels, International Journal of Greenhouse Gas Control 22 (2014) 237–243. https://doi.org/10.1016/j.ijggc.2014.01.008.

[106] P. Ohlemüller, F. Alobaid, A. Gunnarsson, J. Ströhle, B. Epple, Development of a process model for coal chemical looping combustion and validation against 100 kW th tests, Applied Energy 157 (2015) 433–448. https://doi.org/10.1016/j.apenergy.2015.05.088.

[107] M. Kramp, A. Thon, E.-U. Hartge, S. Heinrich, J. Werther, Carbon Stripping - A Critical Process Step in Chemical Looping Combustion of Solid Fuels, Chem. Eng. Technol. 35 (2012) 497–507. https://doi.org/10.1002/ceat.201100438.

[108] C.R. Yörük, F. Normann, F. Johnsson, A. Trikkel, R. Kuusik (Eds.), Applicability of the FB reactor of Aspen Plus for CFB Oxy-fuel Combustion of Estonian Oil Shale: Gas and Solid Hydrodynamics, Mikko Hupa, Turku, 2015.

[109] R. Panday, R. Breault, L.J. Shadle, Dynamic modeling of the circulating fluidized bed riser, Powder Technology 291 (2016) 522–535. https://doi.org/10.1016/j.powtec.2015.12.045.

[110] E.-U. Hartge, C. Klett, J. Werther, Dynamic simulation of the particle size distribution in a circulating fluidized bed combustor, Chemical Engineering Science 62 (2007) 281–293. https://doi.org/10.1016/j.ces.2006.08.067.

[111] A. Abad, P. Gayán, L.F. de Diego, F. García-Labiano, J. Adánez, Fuel reactor modelling in chemical-looping combustion of coal: 1. model formulation, Chemical Engineering Science 87 (2013) 277–293. https://doi.org/10.1016/j.ces.2012.10.006.

[112] A. Coppola, R. Solimene, P. Bareschino, P. Salatino, Mathematical modeling of a two-stage fuel reactor for chemical looping combustion with oxygen uncoupling of solid fuels, Applied Energy 157 (2015) 449–461. https://doi.org/10.1016/j.apenergy.2015.04.052.

[113] A. Abad, J. Adánez, F. García-Labiano, L.F. de Diego, P. Gayán, J. Celaya, Mapping of the range of operational conditions for Cu-, Fe-, and Ni-based oxygen carriers in chemical-looping combustion, Chemical Engineering Science 62 (2007) 533–549. https://doi.org/10.1016/j.ces.2006.09.019.

[114] A. Thon, M. Kramp, E.-U. Hartge, S. Heinrich, J. Werther, Operational experience with a system of coupled fluidized beds for chemical looping combustion of solid fuels using ilmenite as oxygen carrier, Applied Energy 118 (2014) 309–317. https://doi.org/10.1016/j.apenergy.2013.11.023.

[115] J. Fermoso, B. Arias, C. Pevida, M.G. Plaza, F. Rubiera, J.J. Pis, Kinetic models comparison for steam gasification of different nature fuel chars, J Therm Anal Calorim 91 (2008) 779–786. https://doi.org/10.1007/s10973-007-8623-5.

[116] M. Tomaszewicz, G. Łabojko, G. Tomaszewicz, M. Kotyczka-Morańska, The kinetics of CO2 gasification of coal chars, J Therm Anal Calorim 113 (2013) 1327–1335. https://doi.org/10.1007/s10973-013-2961-2.

[117] V. Skorych, M. Dosta, E.-U. Hartge, S. Heinrich, Novel system for dynamic flowsheet simulation of solids processes, Powder Technology 314 (2017) 665–679. https://doi.org/10.1016/j.powtec.2017.01.061.

[118] M. Dosta, S. Heinrich, J. Werther, Fluidized bed spray granulation: Analysis of the system behaviour by means of dynamic flowsheet simulation, Powder Technology 204 (2010) 71–82. https://doi.org/10.1016/j.powtec.2010.07.018.

[119] P. Schlichthaerle, J. Werther, Axial pressure profiles and solids concentration distributions in the CFB bottom zone, Chemical Engineering Science 54 (1999) 5485–5493. https://doi.org/10.1016/S0009-2509(99)00289-4.

[120] J. Werther, J. Wein, Expansion Behavior of Gas Fluidised Beds in the Turbulent Regime, AIChE Symposium Series No. 301 90 (1994) 31–44.

[121] E.R. Gilliland, Fluidised particles, J. F. Davidson and D. Harrison, Cambridge University Press, New York(1963). 155 pages.$6.50, AIChE J. 10 (1964) 783–785. https://doi.org/10.1002/aic.690100503.

[122] J.W. Chew, A. Cahyadi, C.M. Hrenya, R. Karri, R.A. Cocco, Review of entrainment correlations in gas–solid fluidization, Chemical Engineering Journal 260 (2015) 152–171. https://doi.org/10.1016/j.cej.2014.08.086.

[123] S.M. Tasirin, D. Geldart, Entrainment of FCC from fluidized beds — A new correlation for the elutriation rate constants Ki∞*, Powder Technology 95 (1998) 240–247. https://doi.org/10.1016/S0032-5910(97)03343-3.

[124] J.-H. Choi, I.-Y. Chang, D.-W. Shun, C.-K. Yi, J.-E. Son, S.-D. Kim, Correlation on the Particle Entrainment Rate in Gas Fluidized Beds, Ind. Eng. Chem. Res. 38 (1999) 2491–2496. https://doi.org/10.1021/ie980707i.

[125] E. Botsio, P. Basu, Experimental Investigation into the Hydrodynamics of Flow of Solids through a Loop Seal Recycle Chamber, Can. J. Chem. Eng. 83 (2005) 554–558. https://doi.org/10.1002/cjce.5450830319.

[126] J.A. Roberson, C.T. Crowe, Engineering fluid mechanics, 3rd ed., Houghton Mifflin Company, Dallas, 1985.

[127] S.P. Sit, J.R. Grace, Effect of bubble interaction on interphase mass transfer in gas fluidized beds, Chemical Engineering Science 36 (1981) 327–335. https://doi.org/10.1016/0009-2509(81)85012-9.

[128] M. Trefz, E. Muschelknautz, Extended cyclone theory for gas flows with high solids concentrations, Chem. Eng. Technol. 16 (1993) 153–160. https://doi.org/10.1002/ceat.270160303.

[129] K. Redemann, E.-U. Hartge, J. Werther, A particle population balancing model for a circulating fluidized bed combustion system, Powder Technology 191 (2009) 78–90. https://doi.org/10.1016/j.powtec.2008.09.009.

[130] P. Basu, L. Cheng, An Analysis of Loop Seal Operations in a Circulating Fluidized Bed, Chemical Engineering Research and Design 78 (2000) 991–998. https://doi.org/10.1205/026387600528102.

[131] P. Basu, J. Butler, Studies on the operation of loop-seal in circulating fluidized bed boilers, Applied Energy 86 (2009) 1723–1731. https://doi.org/10.1016/j.apenergy.2008.11.024.

[132] L. Cheng, P. Basu, Effect of pressure on loop seal operation for a pressurized circulating fluidized bed, Powder Technology 103 (1999) 203–211. https://doi.org/10.1016/S0032-5910(99)00018-2.

[133] P. Bareschino, R. Solimene, R. Chirone, P. Salatino, Gas and solid flow patterns in the loop-seal of a circulating fluidized bed, Powder Technology 264 (2014) 197–202. https://doi.org/10.1016/j.powtec.2014.05.036.

[134] F. Scala, Fluidized bed gasification of lignite char with CO 2 and H 2 O: A kinetic study, Proceedings of the Combustion Institute 35 (2015) 2839–2846. https://doi.org/10.1016/j.proci.2014.07.009.

[135] W. Zhang, F. Johnsson, B. Leckner, Fluid-dynamic boundary layers in CFB boilers, Chemical Engineering Science 50 (1995) 201–210. https://doi.org/10.1016/0009-2509(94)00222-D.

**Lebenslauf**

| | |
|---|---|
| **Name** | Haus |
| **Vorname** | Johannes |
| **Staatsangehörigkeit** | deutsch |
| **Geburtsdatum** | 24.02.1988 |
| **Geburtsort, -land** | Augsburg, Deutschland |

| | |
|---|---|
| **bis 07.1998** | Grundschule in Zusmarshausen |
| **09.1998 – 06/2007** | Gymnasium in Landsberg/Lech |
| **10.2007 - 06.2008** | Zivildienst in Landsberg Lech |
| **10.2008 - 05.2012** | Studium an der Otto-von-Guericke Universität Magdeburg |
| | Abschluss: Wirtschaftsingenieur Verfahrenstechnik, Bachelor of Science |
| **08.2012 - 06.2014** | Studium an der Chalmers University of Technology, Sweden |
| | Abschluss: Energiesystemtechnik, Master of Science |
| **07.2014 - 07.2018** | Wissenschaftlicher Mitarbeiter, Institut für Feststoffverfahrenstechnik und Partikeltechnologie, Technische Universität Hamburg |
| **10.2018 - heute** | Ingenieur Prozessforschung, BASF SE in Ludwigshafen |

www.ingramcontent.com/pod-product-compliance
Ingram Content Group UK Ltd.
Pitfield, Milton Keynes, MK11 3LW, UK
UKHW061827190726
13853UKWH00009B/2468